D1144067

. . . . . lt Borders

PUBLISHING FOR THE WORLD
125 Years
THE JOHNS HOPKINS UNIVERSITY PRESS

Europe without Borders

Remapping Territory, Citizenship, and Identity in a Transnational Age

EDITED BY

MABEL BEREZIN AND MARTIN SCHAIN

The Johns Hopkins University Press . Baltimore and London

 Published 2003
Printed in the United States of America on acid-free paper
9 8 7 6 5 4 3 2 1

The Johns Hopkins University Press
2715 North Charles Street
Baltimore, Maryland 21218-4363
www.press.jhu.edu

Library of Congress Cataloging-in-Publication Data

Europe without Borders: remapping territory, citizenship, and identity in a transnational age / edited by Mabel Berezin and Martin A. Schain.

p. cm.

Includes bibliographical references and index.

ISBN 0-8018-7436-X (hardcover) — ISBN 0-8018-7437-8 (pbk.)

1. Human geography—Europe. 2. Political culture—Europe. 3. Political geography. 4. Europe—Politics and government—1989– 5. Citizenship—Europe. 6. Group identity—Europe. 7. Europe—Social conditions. 8. European federation. I. Berezin, Mabel. II. Schain, Martin, 1940–

JN12.R458 2004
306.2'094—dc21 2003006212

A catalog record for this book is available from the British Library.

Contents

Preface

During the last few years, space as an analytic category and a context for empirical research has migrated from geography, art, and physics to a range of social science disciplines. ***Europe without Borders*** anchors social science's new attention to ***spatiality*** in the analysis of territory, the politics of European integration. Contemporary Europe affords social scientists a dynamic locale to explore issues of space and politics.

The essays in this volume begin from the premise that territory—the political recalibration of geographical space—matters for a full understanding of societal and political change and conflict. Territory has three dimensions that fuel thick attachments. Territory is social, because, independent of scale, people inhabit it collectively; political, because groups struggle to preserve as well as to enlarge their space; and cultural, because it contains the collective memories of its inhabitants. Territory is mental as well as physical, and its capacity to demarcate social, political, and cultural boundaries makes it the core of public and private identity projects. Territory unites issues of identity and membership in physical space. Territory excludes and includes, and it is intimately connected to issues of membership, from the macro-level of the nation-state to the micro-level of the household.

Nationalism, supra-nationalism, transnationalism, and globalization metaphors of reterritorialization and deterritorialization—underscore an emerging historical paradox with political consequences. Contemporary borders and boundaries mutate and become ever more permeable; yet, territory and territoriality endure. Socially consequential politics—that is, political decisions affecting distributive and membership issues—occur within physically bounded spaces.

Europe today is Europe without borders—a geographical space where territory, membership, and identity are sites of contestation and renegotiation. The redrawing of nation-state borders in post-1989 Eastern Europe coupled with the signing of the Maastricht Treaty in 1992 forced into full view issues and movements that had been bubbling beneath the surface of individual national polities. The multi-dimensionality and plasticity of territory establishes it as central to understanding large- and small-scale social and political transformation.

Territory in all its manifestations — from the physical to the political — serves to organize the eleven analytically rigorous and empirically rich studies in this volume. The contributors to *Europe without Borders* deploy a range of methods — ethnographic, historical, and quantitative — to examine from an interdisciplinary perspective the challenges that reterritorialization and deterritorialization pose to contemporary European politics, culture, and society.

Our analytic commitment to territory and to its relationship to membership and identity does not imply that we seek to reify these concepts. The contributors, individually and collectively, share an understanding of conceptual frames as sensitive to time and space — history and culture. We differ, as the essays in this volume suggest, as to how those concepts become salient and articulate themselves in practice. Similarly, the idea that membership and meaningful participation are central to democratic governance informs our common understanding of political space. We differ as to the character and scale of political space in an era of transnationalism and globalization.

Europe without Borders is divided into four parts. The Introduction frames the volume, arguing that territory is not an arbitrary or idiosyncratic organizing trope. Because territory intersects power, nature, and culture, or force, space, and meaning, it is the fulcrum for all forms of human organization. The interpenetration of territory, capital, and culture made the nation-state and the modern era. Deterritorialization, reterritorialization, and globalization are making the postmodern era.

Part I deals with the spatial and historical characteristics of Europe as a political community. The essays in this section take up the issues of the persistence of the nation-state, the effects of immigration on new definitions of political community, and the contradictions between place and democratic practice which the new Europe renders transparent. Krishan Kumar's essay points out that Europe, old as well as new, has always been a transnational and multicultural arena and describes the contributions of various religious groups, including Muslims, to the historical meaning of Europe. Drawing upon geographical and political theory, J. Nicholas Entrikin elucidates how Europe and European integration underscore the longstanding tension in democratic theory between its espousal of universal norms of justice and the practical fact of governance in particular and limited places. Entrikin argues for a spirit of "cosmopolitan place" which firmly roots individuals in a local context while developing their capacity to empathize across borders. Riva Kastoryano's essay takes up the permeability of European borders. Her empirical focus is the transnational networks that unite immigrant—particularly Muslim—groups in the nation-states of Europe. Given that membership in a nation-state determines the sphere of individual rights within Europe, immigrant networks serve a bridging role that paradoxically increases the strength of national commitment within both the country of origin and the country of migration.

The essays in part II address the issue of spatial rescaling from a formal, historical, and policy-oriented perspective. In their respective essays, the authors examine the social and political consequences of regionalization in Europe and how issues of scale impact upon political organization and culture. Roy Eidelson and Ian Lustick use the techniques of Agent-Based Modeling to simulate the effects of changed territorial borders on identity conflicts. They find that territoriality matters more, rather than less, under conditions of globalization. Roland Axtmann's essay is a historical sociology of state structure which draws out the implications of using the Holy Roman Empire as an analytic model, not a political model, for the process of European integration. Neil Brenner analyzes how successive configurations of urban governance within European cities and city-regions have become sites of both local economic initiative and continual crisis. He identifies "glocalizing state regimes" as European integration's answer to the Fordist city regimes of the 1950s and 1960s.

Part III deals with how ordinary citizens experience "Europe" as a cognitive and cultural construct. Juan Díez-Medrano's essay juxtaposes the elite construction of Europe in the mass media against the popular experience and expectation of a "new Europe." Based on interviews in six substantively selected European cities, Díez-Medrano finds that cultural repertoires based on both the historical memory and lived experience of the old Europe deeply influence visions of, and possibilities for, the new Europe. Levent Soysal's ethnography of Berlin youth culture viewed through the lens of city-based cultural projects shows how public space and performance helps immigrant youth transform an alien urban environment into an integral part of their new transnational identities.

Part IV takes a broad and synthetic approach to the tensions, complexities, and opportunities embedded within new visions of Europe. John Agnew's essay examines the secessionist claims of Italy's Northern League. The League's leader, Umberto Bossi, founded a miniature nation-state, Padania, within Lombardy and the Veneto regions in northern Italy that looks favorably at Europe but unfavorably at fellow Italians—particularly Italians from the south. Agnew's essay develops a political geography of identity formation which places the issues of territorial scale and historical contingency at its center. Craig Calhoun's essay serves as a form of coda to the volume, as it correlates well with the issues raised in the introduction. Calhoun focuses on the formation of a new European public sphere and its relation to a democratic integration of Europe. He argues that a fully democratic Europe is a choice that will very much depend upon how popular interests and public identities merge or clash. An open public sphere will facilitate democratic choice by fostering an inclusive and heterogeneous feeling of solidarity among European citizens.

Europe without Borders began as a one-day workshop held at the New York University Center for European Studies on April 7, 2000. Workshop participants were engaged in research that directly related to the theme of territory, citizenship, and identity in the new Europe. Six of the original workshop participants became part of the completed book project. Neil Brenner, Roy Eidelson, J. Nicholas Entrikin, Riva Kastoryano, and Ian Lustick joined the project after the workshop. We specifically wish to thank Michelle Lamont and Aristide Zolberg for their participation in the workshop, as well as John Glenn who served as discussant at the workshop. Special thanks are due to the staff of the Center for European Studies, who provided the organizational support for the workshop, and the European Union Center of New York, which provided the financial support for participants' travel.

Our approach was informed by a collective hunch that territory in all its dimensions, as well as its modern iteration—the nation-state—would matter more in accounts of European integration and Europeanization than standard accounts in either the social policy or cultural studies literature suggest. *Europe without Borders* is, of course, an academic book written by academics. However, as one of our contributors pointed out, many of us have lived our subject matter. Our contributors live and have lived in transnational spaces—some in multiple spaces. In addition to the American contributors, some of whom may claim hyphenation in one way or other, many of the contributors are transplanted, making even stronger and more counterintuitive our claims about the enduring pull of the national. By accident of intellectual affinity, our contributors are representative of the phenomenon that the book seeks to explore.

The hospitality and resources of several institutions in various transnational spaces contributed to the production of this book. The workshop began as a joint project between the Center for European Studies at New York University and the department of geography at the University of California at Los Angeles, when Berezin was affiliated with both of those institutions. She wishes to thank John Agnew, then-chair of the Geography Department at University of California, Los Angeles, who was an enthusiastic supporter of the idea, and her co-editor, Martin Schain, who gave the funding and institutional support for the initial conference. We collectively thank Henry Tom, our editor at the Johns Hopkins University Press, for his enthusiasm about our project. Henry Tom and his staff, particularly Michael Lonegro, who was the Editorial Assistant on the project, were assiduous about bringing our book to light.

While Martin Schain held the fort on West 10th Street in Manhattan, Berezin took the manuscript in various stages on numerous bicoastal and international airline trips, along the way accumulating a host of debts. The first version of the manuscript was produced at the department of sociology of Stockholm University in Sweden, where she was a visiting professor at various intervals during 2000–2001. Professor Bo Strath's invitation to his economics and politics seminar at the European University Institute in Florence, Italy, helped her to think through her discussion of new versions of European solidarity. The hospitality of Professor Strath and his colleagues, Christian Joerges and Peter Wagner, is much appreciated.

During another stage of manuscript preparation, while Berezin was a fellow at the Hoover Institution at Stanford University, Doug McAdam graciously welcomed her at the Center for Advanced Studies in the Behavioral Sciences in Palo Alto and permitted her to make use of many Center resources. She completed the project in Fall 2002, after joining the sociology department at Cornell University. For his enthusiastic support in making this possible, Berezin thanks Victor Nee. She also thanks Richard Swedberg, who made transnational more than an analytic category and demonstrated the value of border crossing.

In the last analysis, an edited volume such as this rises and falls with the contributors. We could not have wished for a more engaged, enthusiastic, intelligent, and patient group of contributors. *Europe without Borders* is as much theirs as it is ours.

Mabel Berezin
Ithaca, New York

Martin Schain
New York, New York

Europe without Borders

Territory, Emotion, and Identity

Spatial Recalibration in a New Europe

MABEL BEREZIN

The Geographical Moment: ". . . a non-territorial nation"

In July 2000, the International Union of Roma (IUR), more conventionally known as Gypsies, met in Prague. Between 7 million and 8 million Roma live in Europe, particularly in Mediterranean countries and in eastern Europe, where official nationality codes classify them as ethnic minorities. The fall of the Berlin Wall was particularly brutal to Gypsies. Levels of unemployment increased among them and discrimination against them intensified. Before a group of three hundred delegates from thirty-nine countries, Emil Scuka, the secretary general and founder of the first Roma political party in Czechoslovakia, argued that the time had arrived for the global community to recognize the Roma as a cohesive nation—albeit without territory.

Scuka proclaimed: "[I]n contrast to others [European nation-states], the Roma nation is not limited to a territory. It is not a question of a State with precise frontiers, but it is ***a non-territorial nation*** that represents at the same time our specific identity" (emphasis added).[1] If Europe, the geographical space that Roma have inhabited for centuries, is moving toward becoming a "supra-national reality," where "nationality and citizenship do not necessarily coincide"; if hyphenated citizenship yields French-Europeans and German-Europeans as opposed to French and Germans, why, Scuka asks, should there not be Rom-Europeans? Scuka said that the IUR should act as a transterritorial organization, a special genre of NGO to represent Roma at the United Nations. He exhorted the IUR to begin the "unification and codification" of the Roma language—a traditional first step on the path to nationhood.

Scuka was opportunistically seizing the European moment. His logic was impeccable. If the world is global, who is more global than a historically existing transterritorial group? If Europe, as opposed to individual nation-states within Europe, is the point of reference, then why not recognize the Roma as Europeans, given their long-term presence in virtually all of the existing nation-states of Europe? Scuka inadvertently seized the geographical moment in the social sciences as well.[2] Space as an analytic and empirical category has migrated from geography to multiple social science disciplines (Gieryn 2000). Geography, strictly speaking (climate, topography, water supply, and access) inscribes space with its natural particularity; place—the built environment—inscribes space with its cultural particularity (Entrikin 1991; Casey 1997; Tuan 1977). Territory, the political recalibration of geographical space, intersects power, nature, and culture.

From the mid–eighteenth century to the present, some articulation of the sovereign nation-state has embodied modern political territoriality (Poggi 1978). Recent social science argues that increasingly supranational organizational bodies, such as the European Union, have contributed to a "decline" or "crisis" of the nation-state (Graubard 1995; Joppke 1998). Conceptual frames such as globalization, multiculturalism, transnationalism, and nationalism inform discussions of nation-state crisis. Empirical research on citizenship, immigration, and markets of goods and services examines these concepts at the institutional level. Territory and its relation to identity and membership undergirds and unifies empirical and analytic claims.

Territory is implicit in discussions of nationalism and transnationalism—metaphors of reterritorialization and deterritorialization—and points to an emerging historical paradox with political consequences. Communities may be imagined (Anderson 1991); borders and boundaries may mutate. But territory is inescapable. The proliferation of communication in virtual space coupled with a discourse of globalization may generate feelings of infinite possibility, but persons and groups still inhabit real and delimited, if multiple, physical places. Until the Internet rewrites the laws of quantum physics, virtual communities cannot fully replace geographical communities (Calhoun 1992, 1998a). Socially consequential politics — that is, political decisions affecting distributive and membership issues — are tied to physically bounded spaces.

The Rom claim to identity and nationality in the absence of territorial specificity places in bold relief the cultural, social, and political contradictions that are emerging in contemporary Europe. The poverty and persecution that have plagued the Roma for centuries underscore the insecurity of living without the protection of national membership and attendant citizenship rights. Is it possible for an individual or the group to which he or she belongs to claim full enfranchisement if either has an uncertain relation to territory? What is the meaning of national identity in the absence of territory? And how may we speak about membership and identity on a European continent where the territorially given nation-state that defined these issues for at least two hundred years appears to be mutating — if not dissolving?

The fact of European union coupled with the reputed claims of globalization (e.g., Appadurai 1996; Sassen 2000; Kellner 2002) confronts us with three analytic questions. First, is the territorially bounded nation-state simply a product of eighteenth- and nineteenth-century geopolitics and Western cultural hegemony, whose dominance and utility is receding before new institutional structures? Second, is territory an ephemeral and shifting line that may be drawn anywhere or nowhere, depending upon historical context? Third, does the collective experience of living on a territory with a distinct set of cultural and legal norms over time produce thick attachment based on national security, individual enfranchisement, and shared culture that does not easily dissolve — even in the face of what appears to be a new political and economic order? This chapter explores these questions first at the level of theory and second at the empirical level. The section titled "Conceptualization(s)" views the issue of membership and identity in the modern nation-state through the lens of territory; the section titled "A New European Political Project" looks to selected events in contemporary Europe that sometimes affirm and sometimes challenge the nation-state.

■ Conceptualization(s): How Territory Matters

Because territory intersects power, nature, and culture, or force, space, and meaning, it is the fulcrum for all forms of human organization, from the purely social to the purely political. Territory calibrates space and time. Residents experience territorial calibration as propinquity and duration, community and shared history, emotional attachment and identity. In political terms, the modern nation-state is the institutional repository of these relations and emotions (Jusdanis 2001).

The modern nation-state is also the territorial state, and it owes its institutional durability plus its hold on the popular imagination to precisely its intersection of power, nature, and culture. Modern citizenship is the juridical institution that embeds both identity and rights in the territorial nation-state. Even a cursory look at ethnic or cultural groups who, lacking territory in the fullest sense of the concept, are forced to live in the territory of others reveals that a-territoriality and tension are constant companions. This dyad may be as relatively benign as the pursuit of minority rights or as inhumane as the malevolence of ethnic cleansing and genocide—as well as everything in between. The geographical distance from the east coast of Canada, where the Quebecois vigorously pursue their cultural rights, to Kurdistan and Kosovo is long; the analytic distance is short. Minorities, whether they are refugees or asylum seekers, who live on the territory of others are fundamentally homeless unless they obtain legal membership through citizenship independently of what other rights they may claim (Kymlicka 1995).

Territory

Territory derives its intuitive appeal and resilience from its combination of conceptual plasticity and physical specificity. Territory, territoriality, territorial are metaphors of possession that imply actions—acquisition, exclusion, protection. Secondarily, they imply emotions—love, hate, violence. Territory is not reducible to space; nor is it merely a linguistic trope—although it is a compelling one. As a conceptual frame, territory is active and reactive, simultaneously analytic and empirical. As an empirical entity, territory is manifest in a range of organizational forms from the macrolevel of the nation-state to the microlevel of the household—and every other type of spatial configuration that lies between the public and private.

Territories and borders are coterminous. For this reason, territory is never neutral, and it is deeply connected to the exercise of social and political power. The consolidation of power always requires the closing of frontiers—the free spaces of geopolitics. The salience of territory for social science analysis resides in the distinction between its formal properties and the historical formations that it defines. Achieving full analytic purchase on territory requires parsing the distinction between the formal and the historical. Failure to maintain the distinction between the formal and the historical leads to premature proclamations of territory's obsolescence and attenuates its relation to power. As a formal category, territory suggests a behavioral strategy of boundary making that incorporates a range of political forms and organizations; as a historical instance, territory is constitutive of the modern nation-state (Maier 2000).

Disciplinary affiliation tends to influence whether scholarship concentrates on the formal or the historical. Political scientists, particularly specialists in international relations (e.g., Kratochwil 1986–87; Ruggie 1993), have dominated discussions of territory (for a critique, see Agnew 1999). In contrast to scholars who focus on political economy and national security issues, international relations specialists occupied with issues of national sovereignty have tended toward a historical approach to territory (Jackson 1999; Krasner 1999). Independent of the specific analytic focus, international relations theory equates territoriality with the modern state—if not the nation. Globalization, NGOs, and supranational forms of organization, such as the European Union, have forced international relations analysts to reconceptualize territorially defined ideas of sovereignty.

In an influential article, Ruggie (1993) developed a theory of postmodern territoriality that borrows from aesthetic theory to incorporate new forms of political sovereignty. According to Ruggie's account, the ravages of the Black Death in medieval Europe gave birth to the modern territorial single sovereign state. The plague decimated landed elites and forced the Catholic Church into sharing sovereignty with a rising class of entrepreneurs who capitalized on the political and economic opportunities provided by the demise of their aristocratic competitors. Medieval state sovereignty was an exercise in rational politics in the wake of natural disaster. The partnership between the Church and the aristocracy allowed each its own territorial domains. The exigencies of the market eventually made the Church the weaker member in its partnership with the entrepreneurs. By the mid-fifteenth century, this infelicitous partnership (infelicitous from the point of view of the Church) gave rise to a form of national sovereignty that Ruggie describes as the "single" perspective state—a state viewed from the interests of the new political and commercial classes. Single perspective emerged in painting during the same period, and Ruggie argues: "What was true in the visual arts was equally true in politics: political space came to be defined ***as it appeared from a single fixed viewpoint***. The concept of sovereignty, then, was merely the doctrinal counterpart of the application of single-point perspectival forms to the spatial organization of politics" (159).

By the nineteenth century, nationality laws, language, and security became part of the "single perspective" or sovereign state as the territorial landscapes simultaneously became more contested and more in need of solidification. National conscription coupled with the elimination of mercenary soldiers made the nation-state not only a community of culture and citizens but also a community of soldiers—ready to defend and, if necessary, extend the national borders. The "unbundling of territoriality" characterizes the millennial state. Ruggie argues that contemporary sovereignty is shared at multiple nodal points in an international system of power relations that first "unbundles" the relation between sovereignty and territoriality and, second, redesigns the "single perspective" state as a "multi-perspectival polity." The European Union is Ruggie's principal example of this new type of polity. Ansell (forthcoming) and his collaborators have engaged the "unbundling" metaphor and ask whether the "rebundling of territoriality" might more accurately capture contemporary political recalibration. Authority structures that vary in time and space attenuate sovereignty as a political fact and analytic device and effectively recalibrate territory.

Sociologists, concerned with issues of social and political exclusion, have employed ethnographic (most notably, Bourdieu 1985, 1989; Lamont 2000) and historical methods (e.g., Brubaker 1992, on comparative citizenship law; Torpey 2000, on passport legislation) to explore the formal components of boundary making. Anthropologists view territory as environmental and have focused on the economic and cultural production of locality (Appadurai 1996, 178–98; Borneman 1997). Studies of gangs (Gambetta 1993), vendetta (Gould 2000), and neighborhoods (Anderson 1994) implicitly address the social dimensions of territoriality. Massey and Denton's *American Apartheid* (1993) marshals a vast array of statistical data to demonstrate that spatial segregation of African Americans in explicitly black neighborhoods functions as a kind of Gulag that contributes more to diminished opportunities than overt forms of racism in American society.

Geographers offer the most fully realized accounts of territory (Gottmann 1975). Sack's textured analysis of "human territoriality" (1986) elaborates its formal properties. He argues that territoriality is a "human" tendency intrinsically linked to an urge to exercise power. Power, in Sack's view, does not have to be aggressive. Territory, for Sack, is the "control of area" accomplished through the control of access. According to him, "territoriality is a strategy to establish different degrees of access to people, things and relationships" (20). The controlling of access is accomplished in three strategies: first, classification; second, communication; and third, enforcement.[3]

Classification creates categories, such as borders and frontiers (Sahlins 1989; Adelman and Aron 1999), and physically maps the terrain. The development of cartography is an essential tool of political classification (Biggs 1999; Speir 1941). Communication telegraphs borders and boundaries to a public. In the political sphere, treaties adjudicate and exacerbate border disputes (Lustick 1993; Ron 2000). Enforcement is the province of the schools, the police, and the military. National conscription, which began in the nineteenth century, institutionalized the concept of security as a function of the modern state (Finer 1975).

Sack observes that territoriality is an "indispensable means to power at all levels." He argues: "For humans, territoriality is not an instinct or drive, but rather a complex strategy to affect, influence, and control access to people, things, and relationships. . . . Territories are socially [and politically] constructed forms of spatial relations and their effects depend upon who is controlling whom and for what purposes" (1986, 216). Sack's analysis is formal, and his examples range widely in time and space. However, he anchors his concepts in historically sensitive and contextually specific frames that suggest that territoriality might be a byproduct of capitalism and its culturally specific forms of modernity.

Territory and Experience: Time, Space, and Emotion

Territory has four experiential dimensions that fuel thicker attachments than its purely formal components would suggest. Territory is *social* because, independent of scale, persons inhabit it collectively; *political* because groups fight to preserve as well as to enlarge their space; and *cultural* because it contains the collective memories of its inhabitants. Territory is *cognitive* as well as physical, and its capacity to subjectify social, political, and cultural boundaries makes it the core of public and private identity projects. Emotion is a constitutive dimension of territory. The feeling "mine, not yours; ours, not theirs" colors social and political space.[4]

Max Weber's conception of "ethnic group" ([1922] 1978, 389) and "political community" (901–26) provides a starting point for theorizing the experiential dimension of territory. According to Weber, territory is an essential component of political community because it concretizes power relations in space and delimits an area where violence is legitimate.[5] The political and the territorial are fungible — necessary, but not sufficient for creating a political community. Political community is not reducible to either economics (i.e., market activity) or politics (i.e., territorial control); rather it is a form of association that governs social actions among "inhabitants of the territory" who share culture and bonds of solidarity. The political community is composed of "ethnic groups" who may, or may not, have blood ties that draw them together.

The "belief in group affinity" that creates a sense of "ethnic honor" and "sentiments of likeness" (Weber [1922] 1978, 389, 390) is crucial to the formation of political community. In practice, a common language and a monopoly on closure, both territorial and cultural, are the vehicles of honor, sentiment, and community. Weber argues that "joint memories" shape ties that run deeper than the "merely cultural, linguistic or ethnic." Membership in the political community includes the collective expectation that one would be willing to "face death in the group interest" (903). The willingness to sacrifice one's life for one's community based on the belief in a fictive collective past imbues the "political community" with its "enduring emotional foundations" (909).

Following the logic of Weber's argument, a polity that lacks a shared history or a "community of memory" produces politics but lacks shared culture and bonds of loyalty. Members experience political community as duration. Time is as necessary as space for the formation of a thick political community. Abbott (2001) has argued that duration informs all aspects of social life as all social beings experience time qualitatively. Clock time — measures of hours, minutes, seconds — is a quantitative and constructed measure of what we experience qualitatively. Our expectation that day will follow night, that we will rise and perform our daily tasks, is as important as how we block out those activities. Abbott's claims apply in the political as well as the social realm. Political community, like time itself, is ultimately existential and real to the extent that it is there when we need to turn to it. But just as we would not want to toss out all our calendars and clocks, we would not want to eliminate the symbolic, ritualistic, and legal markers and practices of territorial membership. The "territorialization of memory"

(Smith 1996) that is the standard practice of naming places and monuments after the heroes and events of national history, like the hands of the clock, firmly embeds the experience of national political community in the consciousness of territorial inhabitants.

In general, discussions of nationalism have subsumed the cultural dimensions of territoriality. From the moment that Ernest Renan asked in Paris in 1882 his now oft-cited question "What is a Nation?" students of nationalism have grappled with the issue of whether the nation is a quasi-biological and racial entity—a community of attachment among groups with shared bloodline—or a community of affinity among territorial cohabitants. In the last decade, the literature on nationalism has become voluminous.[6] Much of this recent scholarship has focused on putting primordialism to rest. However, in denying the blood connection, scholars have gone excessively over to the constructed view of nationalism. With rare exceptions (Eisenstadt and Giesen 1995; Miller 1995, 23, 24), contemporary constructivist approaches to nationalism that emphasize narratives, which elite and local cultural entrepreneurs fabricate, neglect the experiential dimension of national identity and overlook the durability of national loyalties and attachments. Constructivism has as difficult a time accounting for the recent resurgence of ethnic nationalism as primordialism had in accounting for the multiethnic nation-state.[7]

Calhoun's 1997 account of nationalism that links kinship to the institution of the nation-state bypasses some of the analytic and empirical difficulties of the essentialist versus the constructivist position. He argues that nationalism borrows its linguistic frames from the language of kinship relations in which everyone participates to some degree. Family metaphors are potent tropes and can be used as the basis of building a shared national identity. The physicality of territory, from the family home to the neighborhood to the nation-state, underscores propinquity in space and duration in time. Propinquity and duration generate familiarity or comfort and create a form of cultural and emotional attachment to and identification with the place that an individual or group inhabits. Tuan (1977, 149) reminds us that "[h]uman groups nearly everywhere tend to regard their own homeland as the center of the world" and that this feeling of attachment anchored in place is "not limited to any particular culture and economy" (154).

James, in a classic essay, "The Sentiment of Rationality" (1956), gives a rational explanation for the emotional appeal of comfort. He argues that the "feeling of rationality and the feeling of familiarity are one and the same thing" (78). Novelty, according to James, is a "mental irritant," whereas custom is a mental sedative. He says the core of rationality is that it defines expectation. The expectation of continuity is rational because it is emotionally satisfying.[8] Viewed from the perspective of comfort, emotional attachment to place, what the Germans call *Heimat,* is not an irrational particularism but a rational response to environmental factors.[9] Modern political community, the nation-state, is durable in time and rooted in bounded physical space. As a consequence, its inhabitants experience national membership as a familiar habit and the national landscape as a comfortable place.

The last dimension of territory is the cognitive—the perception of who is one of us and who is not one of us. Propinquity and duration, as in the cultural aspects of territory, figure in this formulation. We see as one of us, as living on our territory, those with whom we have lived for a long time. For example, in 1900 immigrants from eastern and southern Europe were strange and unfamiliar to native-born white Americans. By 2002 they have melded into generic Americans. This classic assimilation paradigm, which may or may not be replicable in other times and places, masks an obvious but overlooked dimension of social life. Persons or groups with whom we are unfamiliar, from whom we do not know what to expect, are perceived to be encroaching on our terrain. Literary critic Elaine Scarry, in an essay on human rights (1999), has pointed out that it is "difficult" to "imagine other people." In view of the difficulty of "imagining others," sustaining democracy requires the development of what Entrikin (this volume) has described as a *mentalité* of "cosmopolitan place."

Territory and Belonging: The Legal Constitution of Identity

In short, territory is congealed identity that embeds relations of social, political, cultural, and cognitive power in physical space. Neither identity nor territory is cast in stone. Territory is identity to the extent that it gives physical place to the iterations of the self or arenas of identity that constitute social, political, and economic life. But identities may change when territorial boundaries change. Problems arise when groups refuse to either surrender or diffuse their identities or when identities are too embedded in ossified structures to change. For example, a person may not think of herself as a spouse after a divorce—an event that usually changes the territorial boundaries of the household. Similarly, as the carnage in the Balkans suggests, persons cannot be Yugoslavians when Yugoslavia is gone, but they may retreat to other forms of territorial identity that may or may not coincide with political mappings.

Identity is the cognitive form that lends transparency to the emotional dimension of territory. Parsing identity from various angles suggests that emotion is its underspecified core. Identity has two dualities that yield different types of analytic questions. Identity is noun and verb, singular and plural. What is it? How do I categorize it? Who am I? Who are we? With whom or what do we categorize ourselves? How are we categorized by others? Personal and political, individual and collective identities differ. "Who am I?" becomes "Who are we?" Who is one of us and who is not? Identity has an ontological and epistemological status (Somers 1994). It describes a state of being as well as a category of social knowledge and classification. Identity is inextricable from the understanding of the self and is central to participation in meaningful patterns of social and political action. Identity suggests similarity (as opposed to the more fashionable focus on difference) and demands acknowledgment of what Taylor terms a "defining community" (1989, 36). Identities involve the recognition of and participation in a web of social relations or communities that envelop the self and through which individuals feel themselves to be identical with others.

Institutions constitute identities.[10] Law provides the mechanisms that support both institutions and the identities that they define. Public identities principally include citizenship and work identities that are institutionally buttressed by the organizations of the modern nation-state and the market. Interest and rationality govern these identities. Private identities are principally kinship-based. Marriage and inheritance law institutionalizes family ties. Cultural identities—religious, regional, ethnic, and lifestyle identities—are more fluid and may be either public or private depending upon historical context. Identity is practical as well as cognitive. Social existence requires an identity embedded in an institution. Persons who assume identities to which they have no institutional connection either are frauds or suffer personality disorders. The homeless are both de-institutionalized and de-identified.

The multiplicity of available identities does not suggest that they carry equal meaning to those who participate in them. Many identities are "contingent" (Connolly 1991, 173), that is, circumstantial. Identities belong to a category that

Taylor (1989, 63) described as "hypergoods" — objects that are of relatively more value to us than others. Individuals or groups experience some identities as "hypergoods" — objects that are of relatively more value to them than others; and they experience some identities as essentially "contingent" (Berezin 1997, 19–30; 1999). Political identities are a type of collective and individual cultural identity that is particularly vulnerable to contingency.[11] Laitin (1998, 24, 31–32) argues that the "conglomerate identities" that emerge in times of national recalibration resemble what rational-choice theorists describe as a "tipping game." Political identities fractionate into local, regional, and national identities (Agnew and Brusa 1999) as well as ideological identities (Cohen 1985; Berezin 1997). Sahlins (1989, 272–76), in his study of national identity on the border between France and Spain in the Pyrenees, draws upon considerable historical evidence to convincingly demonstrate that national identities are fungible and as instrumental as they are cultural. The area that he studied claimed French, Spanish, regional, or even local identities depending upon political exigencies and exogenous events. In an ideal universe, political identities merge emotional attachment and institutional categories. "I am French" and "We are French" would be both ontological and epistemological statements.

Territory as an analytic category inscribes membership and identity in physical space. *Membership* and *citizenship* are equivalent terms in discussions of belonging in modern political organizations. Citizenship as political practice is a modern phenomenon linked in time and space to the formation of national states. Laws governing nationality were part of nineteenth-century European civil codes.[12] Conceptually, citizenship has evolved from a conception of rights attached to persons (Marshall 1964) to a discussion of rules of inclusion (Brubaker 1992), relational processes (Somers 1993), and rights attached to groups (Orloff 1993). When the object of citizenship moves from the individual to the group, rights become articulated as claims. The concept of membership expands to include the cultural, as well as the legal, valorization of the group and stretches the limits of democratic practices (Offe 1998) and institutions (Turner 2001).[13]

Whether one subscribes to revisionist positions that view citizenship as a vehicle for making claims about rights, as opposed to a status with rights attached, two constants hold. First, activists of various persuasions may seek to define citizenship in categorical and collective terms, but it is still granted to individuals (Schmitter 2001). Citizenship still defines the legal relation between an individual and a national state. Second, focusing upon citizenship as a boundary-making device underscores the exclusionary as well as the inclusive aspects of citizenship, but it also attenuates its affective dimensions. Citizenship is more than simply a juridical relationship. It also signals an emotional bond that arouses feelings of national loyalty and belonging in a politically bounded geographical space.

Nation-State Projects: Embedding Identity and Territory in Institutions

Nation-state project is an analytic concept that demarcates a historically specific form of political organization that weds bureaucratic rationality (the infrastructure of the state) to the particularism of peoples and cultures (the nation as community). The term *project* denotes any set of ongoing actions in which collective actors attempt to institutionalize new sets of norms, values, or procedures. *Project* is a felicitous term because it links culture to organization.[14] Citizenship as a body of laws with corresponding institutions permits modern nation-states to locate persons in time (history) and space (culture). As vehicles of political membership, identity, and belonging, modern nation-states move the epistemological (citizenship as category) toward the ontological (citizen as felt identity).

Competition as well as necessity unites identities to institutions. The success or strength of a political identity project depends upon, first, the other identities with which it must compete and, second, the strength of the competing institutions that buttress those identities. A central paradox underlies nation-state projects. On the one hand, without loyal members, that is, citizens who identify with the project, a state will be at a comparative disadvantage in international relations and competition. On the other hand, a state cannot create new identities from whole cloth. The existing identities from which it borrows or appropriates its cultural claims can, unless completely eliminated, at any moment reemerge to undermine the nation-state project. For example, the Catholic Church as institutional and cultural actor has often found itself in conflict with the competing claims of nation-states.

Aside from the legal norms embodied in citizenship and nationality law that juridically tie individuals to the nation-state, nation-state projects require two kinds of activities to create political identities. The first sort of activity is the compulsory participation in institutions that affect all citizens. These activities typically are the military, the schools, and the national language. For this reason, the suppression of regional language and dialect is usually the first item on the nation-state agenda (Weber 1976). Fighting, learning, speaking, as collective actions, create a shared culture of participation. The second activity is consumption—the consumption of national images, words, and symbols in newspapers, art, literature, theater (e.g., Corse 1997; Spillman 1997). Suny's discussion of primordialism as social construct (2001) argues that the process of repeated exposure to identity-forming institutions, or production and consumption activities, yield a form of national attachment that has a genuine emotional base to it.

The territorially defined nation-state project was not a seamless effort. The development of the nation-state was contingent, conflictual, and contested. Its emergence varied as to time and space—history and culture. Between what historian Eric Hobsbawm has called the "long nineteenth century" and the "short twentieth century," three state-centered political projects in Europe designed variations on the nation-state form. These projects were, first, the nineteenth-century nation-state consolidation projects; second, the conflictual projects of the early twentieth century, of which Fascism and Nazism were examples; and third, the postwar rebuilding period, including the cold war, the Marshall Plan, and the original Common Market. National and international events that occurred during each of these periods suggest that it took hard cultural institutional work in the form of national education policy, language consolidation, the construction of museums and monuments, and the use of music (not to mention war) to consolidate identities around territorially bounded bureaucratic forms—to make citizens of modern nation-states. The European Union is arguably the fourth variation on the nation-state theme.[15] We can now turn our attention to how the issues discussed so far may play themselves out in an emerging Euro-polity.

A New European Political Project: Responses to Territorial Recalibration

The European Union, consolidated in the Maastricht Treaty in 1992, is a supranational political project that transforms Europe into a geographical space where territory, membership, and identity are once again sites of contestation and renegotiation. European integration is the principal process of institutional change at the core of the new Europe.[16] As its creators imagined it, European integration was an economic mechanism to permit goods, persons, services, and capital to flow freely across borders. The process of integration dates to 1950, when French foreign minister Robert Schuman called for a common market in coal, iron, and steel. The European Coal and Steel Community was the first step in a process that eventually led to a series of treaties, beginning with the Treaty of Rome in 1957 and culminating in the Maastricht Treaty of 1992.[17] What began as a trading and tariff organization with 6 members has evolved into a 15-member transnational organization with the possibility of arriving at 27 members as the Union begins to admit nation-states from the former Eastern Europe.

What does *European Union* mean today? In the last decade, a public discourse that focuses on European citizenship and the drafting of a European constitution has emerged, and if it has not replaced, at least it has attenuated the economic mission of the Union.[18] But what can citizenship mean in a supranational body whose members have no direct voice in governance? The "democratic deficit" and solutions to it have been a persistent focus of scholarly attention (most recently, Scharpf 1999; Schmitter 2000). *Democratic deficit,* referring to the fact that many regulatory decisions that affect the daily lives of citizens of individual nation-states are made in Brussels without their explicit consent, is, in practice, a relatively abstract and remote concept to the ordinary person. The deficit becomes real when something like mad cow disease infects European cattle, stops the consumption of beef, and forces individuals to change their daily dietary habits.

European integration challenges the prerogatives of territoriality and by extension disequilibrates the existing mix of national culture and legal norms. By threatening to make the national space unfamiliar to many citizens, it opens a space for contestation as well as positive change. The remainder of this chapter takes up four responses to territorial recalibration in the cultural and social field: identity, populism, solidarity, and postnationalism.

European Popular Identity—A New Political Community?

In 1860, after the legal unification of Italy, Massimo D'Azeglio said, "Having made Italy, we must now make Italians." Today, D'Azeglio's oft-cited injunction to the Italian Senate could just as easily begin a session of the European Parliament—"Having remade Europe, we must now make Europeans." A united Europe in its postwar incarnation was restricted to a common market. Even in 1992, economics, not culture, was principally on the mind of Jacques Delors. In the last decade, the question of a European identity and European citizenship has begun to dominate the discourse of European political and intellectual elites (see, e.g., Delanty 1995; Kastoryano 1998; Hedetoft 1999; Strath 2000; Schnapper 2002; Risse 2001; Borneman and Fowler 1997).

The "old" Europe of territorially bounded nation-states—the by-products of wars, revolutions, and social conflicts—evolved as a political, economic, and cultural project. A new European identity is arguably an afterthought. The raw material for a European political community in the Weberian sense, outlined in the section of this chapter titled "Conceptualization(s)," is flawed on two counts. First, Europe as a political space is territorially ambiguous.[19] Regulatory decrees are trans-European. Membership is nation-state-based. Only individual member states, not the European community, may bestow citizenship. The ability to work across national borders—one of the attractions of the EU for the educated and upwardly mobile middle classes—frequently bogs down in a mass of red tape that defies the rational language of the Schengen Accords (Romero 1990; Daley 2001). Second, Europe as a cultural space lacks "affectivity"—that is, emotional attachment (Weiler 1999b, 329). Old European nation-states, as Weber argues, crafted a fiction of shared culture and history from a widely diffused community of popular memory. Europe has no common popular civic space or cultural past from which to forge an identity except for memories of war—and usually war between member states (Mann 1998; Laurent 2003).

As Darnton (2002) and Kumar (this volume) remind us, European identity per se is not new. A shared high culture among the university-educated—usually European men—who spoke and read in three national languages, usually English, German, and French, in addition to their own, was the Procrustean bed of the old European identity. Exclusive social and professional networks forged one part of the old European identity. Nineteenth-century innovations such as mass schooling and conscription helped to foster bonds of national solidarity among workers and members of the lower middle classes (Hobsbawm 1983; Weber 1976). Post-Maastricht European identity claims to be popular and inclusive. In contrast to old European identity, new European identity is a product of political demand rather than social contingency. The "unbundling" of nation-state sovereignty and the logistical problems that it brings to ordinary citizens are as likely to strengthen existing national identities as they are to generate a feeling of common Europeanness (Wallace 1999; Berezin 1999, 2000).

Bureaucrats in Brussels who seek to turn Germans, the French, and Italians into Europeans face the twin obstacles of a contradictory legal framework and thin cultural demand (Deflem and Pampel 1996). The framing of a European constitution aims to adjudicate legal issues.[20] As I argued earlier in this chapter, a legal framework is necessary but not sufficient to create either a nation-state or a European identity. Modern nation-states were territorially bounded identity "projects" that embedded identities in institutions.[21] The European Parliament and Council seeking to create a trans-European cultural infrastructure have placed identity institutions from education to tourism on their agenda. Many of the identity institutions that are restricted to the cultural sphere remain either partially developed or in the planning stage (Laitin 2000).

Commerce and public communication embedded in European monetary and language policy are the two identity-producing activities that no inhabitant of European space may avoid. National currencies and national languages attach persons to territories and collectivities, define membership criteria, and circumscribe social and economic opportunities (Berezin 2000; Risse et al. 1999; Verdun 1999). Learning a new currency, like learning a new language, is a practical as well as a psychological intrusion. Habituation is at the core of both processes. In 1989 the Council of Europe voted to establish the European Monetary Union (EMU), which exchanged national currencies, the lira, the franc, the mark, for a single European currency—the euro. Between January 2000 and January 2002, the euro was phased in as the currency of those member nations who had voted to participate in the euro-zone. Brussels marketed the euro as a vehicle of European identity and citizenship (European Commission 1998).

The European press marked the cultural as well as commercial significance of the euro when it bade farewell to national currencies on New Year's Eve 2001. The French *Liberation* sighed, "Twelve Countries Take Out Their Handkerchiefs" ("Douze pays" 2001). A front-page editorial in the Italian *La Repubblica* proclaimed, "An Act of Faith Becomes a Reality!" It asserted that the euro would usher in a new political epoch: "The euro, in reality, is the true political act that closes the history of the 20th century, of its tragedies and its divisions, and founds at the same time the new century of Europe" (Mauro 2001). In the company of nostalgia, high hopes, and some expectations to the contrary, the introduction of the euro proceeded smoothly with little popular resistance.

In July 2002, when the euro first outpaced the dollar on world currency markets, its performance seemed to generate a sense of European identity and pride. Europeans bonded over the euro, not because they had a common currency in their pockets, but because they had, at least for a brief moment, done better than the United States, which they traditionally viewed as a competitor. The *New York Times* reported that the advance of the euro gave Europeans a psychological edge: "[A]s a collection of nation-states that have different views about how much power to cede to Brussels, the union has viewed the euro — like the common European passport — as an important symbol of the reality of this experiment in shared sovereignty. And that makes the moment of its breakthrough especially significant to Europeans" (Erlanger 2002).

Linguistic identity, a core feature of past projects of territorial consolidation, poses a more intractable problem than currency. Unlike the euro, a single European language cannot simply be legislated into being. What language will citizens speak in a transterritorial Europe, where multiple languages demand recognition if not domination? The European Union has 15 member states representing 13 official member languages. More than 30 local and regional languages are spoken in the same 15 member states. As the EU expands, the language problem threatens to expand with it. But what precisely is the language problem? The 1957 Treaty of Rome guaranteed the integrity of the national languages of its signatories. This principle of linguistic integrity has been upheld through succeeding treaties. Currently, all official documents, plus parliamentary sessions, must be translated into all the languages. Translation services represent a sizable portion of EU budgets (Kraus 2000).

The EU has considered several alternatives — all of which lack plausibility in varying degrees — from returning to Latin, the official language of medieval European politics, to resorting to a form of European Esperanto. The French have refused to support the one concrete policy proposal on language — European initiative to support the revitalization of regional languages. National chauvinism aside, residents of a transterritorial Europe will find it difficult to limit themselves linguistically to their native tongues. Laitin (1997) argues that to solve the European language dilemma, Europeans will need to speak two languages, plus or minus one, depending upon the degree of diffusion of their national language. One cannot escape the irony that English may emerge as the lingua franca of the new Europe.

Reaffirming the Old Political Community: The Populist Response to Europe

The tangle of interpretation and administration that besets the European Union in the fullest sense of the term (juridical, social, and cultural) suggests that a transnational polity with a loyal and attached citizenry—a citizenry that identifies itself as European—lies in the distant future. It also points to an unpleasant underside of union. Political aggregation upward yields social disaggregation downward, and downward disaggregation has the potential to create political and cultural disruption and conflict. Euro-mobilization is an evolving phenomenon. Imig and Tarrow (2001) and their collaborators have explored protest events in new European political space. Multilevel mobilization at the subnational, cross-national level, as well as at the European level, differentiates Europeanized contention from previous forms of contention (Tarrow forthcoming).

To date, the weight of protest against Europe has been in the parliamentary rather than the extra-parliamentary sphere. The emergence, and in some instances the electoral success, of right-wing populist parties has been one "disruptive" response to the expanded process of European integration.[22] In the past, discussions of national sovereignty and identity, always central to European unity, were restricted to the governing strata. The electoral success of formerly fringe parties such as Jean Marie Le Pen's National Front in France, Jorg Haider's Freedom Party in Austria, and various fringe parties in Switzerland, Belgium, Denmark, and the Netherlands have catapulted the European Union into the public sphere and made the national political community a subject of popular debate.[23]

The National Front provides the best illustration of these trans-European populist tendencies and serves as a yardstick against which to measure other parties. The National Front has been on the French political scene since 1972. Widely viewed as a fringe ultranationalist party, the Front began, between 1983 and 1986, to attract significant portions of the electorate. In the 1988 presidential elections, Le Pen received a surprising 14.4 percent of the vote (Schain 1987). Scholars and pundits alike attributed Le Pen's success to his belligerent stand against immigration and his racist ravings in a country that some segments of the population per-

ceived as overcome with immigrants. In response, French politicians concerned with the national commitment to republicanism and Frenchness began their own effort to restrict immigration and to reform the Nationality Code (Schain 1996; Weil 2001; Feldblum 1999).

The respectable center-right's co-optation of his positions did not deter Le Pen. His party won three mayoral elections in 1995. Success in the mayoral elections encouraged Le Pen to launch a full-scale mobilization for the regional elections of 1998, where again the Front won a surprising 15.3 percent of the vote. The Front attacked the European Union as a "utopian scheme of Eurocrats," in contrast to the National Front, "the Party of France."[24] Just as the French state had absorbed the immigration issue, it also managed to absorb and diffuse the "Europe" issue. Alarmed by the Front's success in the 1998 regional elections, a center-right coalition headed by Charles Pasqua (who was also the author of the new Nationality Code) managed to lure its less fanatical supporters. The center-right had famously argued, "The Front poses good questions . . . it simply gives bad answers!" The emergence of a center-right alternative, coupled with an internal split within the National Front leadership, sent the party to a crushing defeat in the European elections of June 1999.

Le Pen and the National Front, while they did not disappear from the political landscape, appeared to have lost their political clout after the 1999 elections. In spring 2002 a structural oddity in the French electoral system, a two-round ballot process in the presidential elections, catapulted Le Pen to the forefront of French politics once again. Le Pen came in a surprising second to Jacques Chirac in the runoff for the presidency of France. The lackluster campaigns that front-runners Chirac and Lionel Jospin ran for the presidency contributed to Le Pen's success. First, many small parties competed in the election, splitting the left and center-right vote; second, a bored and disaffected populace stayed away in droves. The election had the highest rate of abstention, 28 percent, in any post-1958 French election.

The idea, as much as the reality, that Le Pen would be a serious contender for the presidency of the French Republic sent shock waves throughout France, "le choc" as the French referred to it, and the international community. The 2002 presidential campaign between incumbent president Jacques Chirac and Socialist prime minister Lionel Jospin was lackluster—with neither candidate garnering strong support. The specter of Le Pen as president of France led to many political oddities. Socialists and leftists of all stripes ended up voting for the center-right, even if some did it with clothespins on their noses. Chirac, who had been the center of a corruption scandal and was popularly referred to as a crook, was returned to office with a vote of 82 percent — the highest ever in modern French history. Spontaneous mobilization occurred in the streets of Paris and all over France. On May Day, over a million people took to the streets of Paris — the only demonstration that had been larger was the 2 million that emerged in the streets when Paris was liberated from the Nazis in 1944. Chirac's campaign slogan was "Yes to the Republic," in contrast to Le Pen's more nationalist-oriented depiction of himself the party of France. In the end, the Republic — assimilationist, democratic, and French — not European — won.

The rise and fall and rise and fall of the National Front brought the issue of national identity and Europe to the front of the French political agenda. In Austria, Jorg Haider's Freedom Party brought the issue of national sovereignty to the forefront of European public debate. When Haider's party achieved enough votes to become part of a governing coalition, the European political community responded with alarm and moral outrage. In the past, Haider had made violent statements against immigrants and had defended the Nazi persecution of the Jews. He justified his defense of the Holocaust by saying that members of his parents' generation were not to blame for Nazi atrocities. To ensure that his party would become part of the government and that new elections would not be called, he quickly "apologized" for past mistakes and immediately stepped down from his position as party head. Haider retreated to his base in the border province of Carinthia.

Despite Haider's retreat, there were international calls for Austria to redo the election. The European Parliament, which had no authority in this matter, placed sanctions on Austria as a member of the EU. Austrians perceived the EU sanctions as a violation of national sovereignty. A democratic national election had brought the Freedom Party into the government. On one level, the sanctions backfired: first, they increased the popularity of the Freedom Party among Austrians who resented "foreign" interference in their national election; and second, they raised the issue of an organization without a democratic mandate attempting to undercut the authority of a national state. Since the sanctions had no juridical force and were, in practice, rather pallid, amounting to little more than snubs to Austrian diplomats and ministers at international forums, a cynical interpretation might be that the Parliament might have perceived the Austrian elections as an opportunity to show that even if it had a "democratic deficit," it did not suffer from a moral deficit. By May of 2000, members were already speaking of lifting the sanctions.[25]

A Neoliberal Political Community: The "Unbundling" of Institutionalized European Social Solidarity

Very public events, such as the French and Austrian elections, are more often than not transient and deflect attention from more fundamental social changes. Territorial recalibration brings with it conceptual recalibration. The "unbundling of territoriality," to return to Ruggie's phrase, also has the capacity to "unbundle" collective *mentalité*. Europe, in contrast to the United States, always had a more finely honed sense of the social. European nation-states were more or less solidaristic. Solidarity, in a Durkheimian sense (simply put, a tacit acknowledgement that the whole functioned better if the parts were in harmony), informed national policy decisions.[26] When membership or citizenship was attached to tightly bounded territorial nation-states, a vision of society based on solidarity and mutuality was possible to legislate.

The European Union's new Charter of Fundamental Rights and the opposition to it are the first tangible signs of the "unbundling" of what had become an institutionalized form of social solidarity. Participants in a February 1999 meeting of the European Commission on Employment, Industrial Relations, and Social Affairs decided to design a document that outlined fundamental rights in the European Union (European Commission 1999). Union expansion to the countries of the former Eastern Europe, demographic changes, globalization, and even rising crime rates (the French National Front campaigned on the issue of "sécurité") suggested that the time was propitious to come up with a set of common European principles. On the surface, affirmation of rights is laudable. From the perspective of politics, the action was puzzling. The 1996 revision of the European Social Charter, originally passed in 1961, had already accounted for changes in the structural positions of women and the influx of immigrants (Harris and Darcy 2001).

A sequential reading of the European Social Charters of 1961 and 1996, followed by the Charter of Fundamental Rights, places in bold relief how much, and in how short a time, the political culture of Europe has changed. The 1961 charter was written with the view that full employment was a goal, the International Labor Organization was a party to the charter, and it firmly supported the idea that the family was the basic unit of society. It was, socialist or not, a collectivist document. The 1996 amendments kept the basic spirit of the 1961 charter while bringing it up to date with contemporary issues such as gender discrimination, informed consent, the right to housing, and the rights of the disabled. Work and family, labor and community were at the core of the Social Charter in both its iterations.

Read against these two previous documents, the Charter of Fundamental Rights represents a striking departure from an earlier political culture. The sense of the social is absent from the new charter, and it is replaced with an affirmation of the individual. Rights, as I discussed in the earlier section on citizenship, are always attached to individuals, but in the past they also implied and conveyed entitlement. The new charter is individualist because it replaces entitlement with a neoliberal version of freedom that shifts responsibility from the polity to the person. A few examples from the charter serve to illustrate this point (Charter of Fundamental Rights of the European Union 2000). The individual has the right to freely seek employment; society is not responsible for creating jobs (Art. 15). Women have the right to be free of harassment and discrimination—once they have found their market niche (Arts. 21, 23). The family as the basic unit of society has been replaced by the guarantee that the family has an absolute right to privacy (Art. 7).

The European Council unveiled the Charter of Fundamental Rights at its biannual meeting in December 2000 at Nice, France. A left-wing group, ATTAC, mobilized fifty thousand persons to travel to Nice and engage in three days of public protest against the charter. Bernard Cassen, editor of *Le Monde Diplomatique,* and an assortment of trade unionists, intellectuals, and human rights activists founded ATTAC (Action pour une taxe Tobin d'aide aux citoyens) in Paris in June 1998.[27] ATTAC campaigned against the charter with the slogan "Another Europe Is Possible." Proclaiming that the European Union had become a "motor of liberal globalization," ATTAC argued that the new charter was fundamentally antilabor, antisocial, and antinational. The weakening of social rights was among ATTAC's principal concerns—specifically Article 15, which only ensures the right to look for work. ATTAC was not alone in its antipathy to the charter. Although it was signed unanimously, representatives of the member states were not united in their enthusiasm.

From the vantage point of the twenty-first century, it is easy to forget that solidarity in the old Europe was not only laborist. It was embedded in the collective mentality of long-defunct aristocracies as well as in bourgeois networks of family capital and national firms. Global capital depersonalizes capital transactions. In the United States, with its individualist ethic, the desocialization of capital is hardly noticed—until its economic effects are felt. In Europe, desocializing capital sends shock waves through the political and social system. From that perspective, the new charter protects individuals against the abuses that are constitutive of unbridled market forces. But because this strategy focuses upon the individual rather than the collective, it represents a distinct rupture with past practices in Europe and a challenge to—if not the complete end of—European versions of social solidarity.

Transcending Political Community: Postnational Discourse and Structural Imperatives

Populism is a negative response to Europeanization and territorial recalibration and a claim to maintain the old national political space. Postnationalism and multiculturalism as the foci of public discourse articulate with the positive potentialities of Europeanization. The increasing presence of immigrants in the territories of established European nation-states, particularly but not exclusively France and Germany, not only has been an issue with right-wing parties but has also pushed the discussion of postnationalism and multiculturalism to the forefront of political debate as well as social science research.[28]

Soysal's *Limits of Citizenship* (1994) is the most frequently cited work on the subject. Based on her study of immigrant associations in six European nation-states, Soysal identifies four types of "incorporation regimes." She concludes, based upon her analysis of these regimes and their iterations in diverse territorial states, that a new form of citizenship has emerged that decouples territory from legal membership. She labels this "post-national citizenship" and argues that it points to a new form of transterritorial membership that is based upon human rights—the rights of persons as persons rather than persons as members of nation-states.

Scholars have contested the "postnational" argument—Soysal's variant as well as other articulations of it (e.g., Jacobson 1996; Tambini 2001). Postnationalism as theory is based on a paradox that squares poorly with political reality (Eder and Giesen 2001). Postnationalism upholds the autonomy of national cultural difference at the expense of political membership. By privileging culture and nature, nationality and humanity, over territorially based institutional ties, postnationalism as concept leaves itself open to criticism that it is utopian and that, in practice, it may actually threaten the legal rights of migrants. The paradox at the core of the postnational argument is also present in discussions of multiculturalism (Joppke and Lukes 1999). In the European context, with increasingly public and mobilized immigrants in national territories, multiculturalism is more than simply the "United Colors of Benetton" in the political sphere.

L'affaire des foulards, as it became known in France, is a celebrated example of the contradictions of multiculturalism (Beriss 1990; Feldblum 1999, 129–45). In 1989 the principal of a French high school expelled three Muslim girls from school for wearing the traditional headscarf or veil to class. He argued that the wearing of the veil in a public school violated the republican principle of separation of church and state. The debate over the veil catapulted the French public sphere into a state of frenzy for months. Lionel Jospin, then minister of education, delivered what became the final ruling—children should be allowed in schools. Religious insignia were irrelevant as long as the children did not proselytize for their religion. The final solution pleased no one—except perhaps the girls who were expelled. France's first multicultural conflict was a harbinger of more to come throughout Europe as national cultural practices collided in bounded territorial spaces.

Despite the scholarly discourse on postnationalism and the political fact of European integration, compelling counter arguments exist from a purely structuralist perspective (e.g., Mann 1997; Evans 1997) that suggest that the territorially defined nation-state is hardly withering away. The European Charter of Fundamental Rights (2000) defers claims over rights to "national laws and practices." Examples from recent empirical research suggest that this deference is not merely normative but a recognized component of political practice. Koopmans and Statham (1999) tested the postnational hypothesis by examining immigrant claims in Britain and Germany. They found that minorities structure their claims in the language of citizenship and rights prevailing in the national territory where they find themselves and not in terms of the national identities and cultural practices of their homeland. Bhabha (1999) demonstrates, using data from cases before the European Court of Justice, that residents of a territory who are not legally incorporated members of the territory, that is, citizens, have little recourse to the full array of constitutionally protected rights. Many of her examples focus on marriage. Citizens of nonmember states, even if married to naturalized citizens, face the threat of deportation.

Bhabha's and Koopmans and Statham's research underscores the point that a European is only European, as defined by the European Union, if he or she is a citizen of one of the member states. Legally, transnationality within Europe is a tightly bounded concept. Indeed, the juridical evidence makes postnationalism appear moot. The continuing hegemony of the nation-state, even in the presence of an expanding European Union, suggests why *transnational* is a better descriptor of the contemporary European political culture than *postnational*. *Transnational* captures the hybrid potential implicit in the postnational without attenuating the difficulties of a rapidly diversifying Europe.

Demographic necessity may render the postnational discussion moot. After borders, population is the core of any territorial unit. Robust natality is essential to any nation-state project — whether it be democratic (Watkins 1991) or state socialist (Gal and Kligman 2000). Until recently, European politicians, as well as social scientists, have viewed immigration and fertility as two separate issues (Livi-Bacci 2000, 163–89). These two structural variables are increasingly affecting the political development of the European Union and the labor market structures of member states. In the past twenty-five years, immigration rates have soared in the core of Europe. There has been a corresponding decline in birthrates among the native-born. In Germany and Italy, birthrates have dropped below what demographers label as "replacement" (United Nations 2000).

Birthrates are declining independently of the level of social support for maternity — suggesting that rather than demographic decline there is a long-term secular trend toward "demographic refusal" among the nonimmigrant populations (Teitelbaum and Winter 1998; Bruni 2002). The European Right was the first group to sound the alarm that only the immigrants were reproducing. Demographic policies from which European politicians had distanced themselves because they evoked the eugenics and genocidal policies of the 1920s and 1930s have begun to appear attractive to center and left politicians. Declining birthrates combined with an aging population have begun to threaten the economic foundations of the European Welfare State — a traditional focus of the European Left (Esping-Andersen 1997).

Nation-states need replacement populations to sustain the labor force as well as to contribute to the pension funds that support their aging populations. Without replacement populations, the entire European social welfare system, already under siege, is in danger of collapsing. Ironically, immigrants and their communities are an important factor in salvaging the welfare state. Immigrants to established European nation-states, for cultural and economic reasons, reproduce at a higher rate than the native-born. Full incorporation of immigrants on the territory, that is, via citizenship, will allow their participation in the types of labor markets that typically provide the tax base for the welfare state. Like many things in the new Europe, the answer is not in yet on the question of demographic decline. However, there is graffiti, if not handwriting, on the wall. By the year 2050, according to United Nations' projections (2000, 25), over 60 percent of the populations of France, Germany, and Italy will be descendants of non-native-born persons. By numbers alone, immigrants are the future of Europe — and whatever chance remains to preserve the old Europe of social welfare solidarity resides with them.

■ Conclusion

Territory, and its modern iteration the nation-state, calibrates power, nature, and culture in bounded physical space. Theorizing the experiential dimension of the nation-state through the lens of territory underscores its durability as a political form. Political, social, and cultural mechanisms support this durability and create powerful ties of emotion, identity, and practice. Propinquity rooted in the natural and built environment, coupled with legally guaranteed institutional arrangements, produces and reproduces the nation-state. Shared history and national memory are by-products of nation-state durability.

Territory as a formal category of analysis, and a historical entity is an important organizing trope for interpreting political change in contemporary Europe. In volume 1 of *Das Kapital,* Karl Marx famously argued that a commodity was a wondrous and multifaceted entity because it contained congealed human labor power. Marx's metaphor applies equally to territory as a repository of congealed identity, emotion, and power. The interpenetration of territory and capital created the nation-state and ushered in the modern era. Forms of deterritorialization and reterritorialization, from globalization to Europeanization, are producing the postmodern era.

Spatial recalibration in Europe presents opportunities as well as challenges. Territory is durable but not eternally fixed. Opportunities will emerge that lie somewhere between populist pessimism and postnational optimism. Whether one accepts strong or weak versions of the various "post" hypotheses, in a world where supranational organizations such as the European Union challenge nation-state projects from above, and ethnic and regional conflicts challenge those projects from below, territory, in all its dimensions, matters.

NOTES

Thoughtful comments from Richard Biernacki, Riva Kastoryano, Doug McAdam, Richard Swedberg, Sidney Tarrow, and two anonymous reviewers helped me to revise an initial version of this chapter.

1. "Interview—Emil Scuka," *Liberation* (Paris) 25 July 2000, 9.

2. Examples of recent scholarship in which geography is salient to the analysis include Baldwin 1999, on the "geo-epidemiology of disease"; Mukerji 1997, on the relation between garden design and French state formation; and Diamond 1997, on human development.

3. Sack's arguments will resonate with readers who are familiar with Pierre Bourdieu's oeuvre. In contrast to Sack, Bourdieu views space as more of a metaphor than a physical place, and this renders its relation to power rather slippery.

4. Territory as cognitive and emotional experience has received relatively little sustained scholarly attention. For innovative approaches to cognition, see Wagner-Pacifici 2001, on "narratives of surrender," and Zerubavel 1992, on mental mapping. Territoriality as an emotion is often analyzed negatively in terms of violence (Brubaker and Laitin 1998) or "modern 'barbarity'" (Offe 1996).

5. While Weber categorically links territory to political community, he does not specify that the political community must take the form of the modern nation-state. See Rokkan 1975 for a theory of the nation-state and territoriality.

6. Smith (2000) summarizes much of this literature. His evolving discussion (65–71) of nationhood as a relation between *ethnie,* a human population sharing a homeland and myth of common culture, and nation, the ideological appropriation of those myths, owes much to Weber in its formulation.

7. The reliance on the term *nationalism* has led to a distinction between ethnic and civic nationalism (i.e., good and bad nationalism). *Patriotism* (Viroli 1995; Nussbaum

1996) is a more useful concept because it recasts loyalty and attachment in more positive terms.

8. For a fuller articulation of this position and its relation to politics, see Berezin 2002.

9. For a critique of the uncritical acceptance of this view, see Calhoun 1999.

10. While there is a large literature on both identity and institutions, scholars have never explicitly linked the two. Friedland and Alford, in their discussion of multiple institutional affiliations (1991, 247–56) address some of these issues.

11. I use the term *political* in the sense of identifying with a polity. I choose this term over the more popular *national* because *national* frequently confuses the object of identification. In short, the term *national identity* conflates juridical attachment and loyalty to a nation-state with excesses that are sometimes, but not always, constitutive of nationalism.

12. On the development of nationality laws in various European nation-states, see the collection of essays in Hansen and Weil 2001.

13. For a summary of different iterations of the term *rights,* see Jones 1999.

14. I use the term *project* (Berezin 1997) to speak about the Italian Fascist state; Fligstein and Mara-Drita (1996) use a variation of the term to discuss the formation of the Single Market Policy in Europe.

15. Goldstein (2001) takes up the issue of "federal sovereignty" in comparative perspective.

16. In a field densely populated by scholarly and journalistic studies, Moravcsik (1998), Ross (1995), and Milward (2000) present rigorous full-length studies that serve as introductions to the issues involved in integration.

17. See Sweet, Fligstein, and Sandholtz 2001 for a discussion of this process. Although for all intents and purposes the Maastricht Treaty signals the completion of the European Union, it was amended twice—at Amsterdam in 1997 and at Nice in 2001.

18. Bruneteau (2000) argues that European Union, or the idea of it, was a linchpin of instrumental and contingent national political goals and not a normative vision of European solidarity.

19. For concise technical discussions of the legal issues involved, see Wouters 2000, on national constitutions, and Davis 2002, on citizenship law.

20. See the exchange between Grimm (1997) and Habermas (1997).

21. My approach differs from a standard constructivist argument, as it emphasizes the capacity of state actors to use institutions to connect with publics as audiences. Christiansen, Jorgensen, and Wiener (1999) offer a constructivist approach.

22. The relation between xenophobia and immigration policy has dominated issue-oriented studies of the European Right (e.g., Schain 1996).

23. European left and right populism is an evolving political phenomenon whose meaning and extent is still unclear. Much of the narrative here is taken from my own ongoing research. For a recent summary, see Eatwell 2000. Kitschelt 1995 is the standard empirical work on the new European Right.

24. I describe these issues at greater length in Berezin 1999.

25. I base this discussion upon a broad reading of the European and American press at the time. See, particularly, Judt 2000; R. Cohen 2000.

26. There is a large literature on the development of the European Welfare State—much of which attributes its development to various forms of popular struggle, from workers' movements to women's movements. It is beyond the scope of this chapter to examine that literature. My basic point is that there was, for a large array of reasons, a different, more community-based vision of society in Europe than in the United States that contributed to the relative success of those movements in Europe and their relative failure in the United States. Although he does not take the charter into account, Rosanvallon's argument (2000) resonates with mine.

27. See Keck and Sikkink 1998 for a discussion of organizations such as ATTAC.

28. Examples include Money 1999, on France and Britain; Kastoryano 2002, on France and Germany; Pred 2000, on Sweden; Wieviorka 1992, on France; Baldwin-Edwards and Schain 1994, on multiple nation-states.

PART I

Political Community in a New Europe

ONE

The Idea of Europe

Cultural Legacies, Transnational Imaginings, and the Nation-State

KRISHAN KUMAR

> [W]hat constitutes the essence of European history remains contentious and unclear. . . . Historians cannot be used to provide a simply usable pedigree for the present membership of the European Community, much less to suppose that it represents the inevitable climax of all European history.
>
> KEITH ROBBINS

> I would defend myself, weapons in hand, if they tried to prevent our listening to Croatian sermons in church or speaking Croatian in a café, but I do not want my children going to a Croatian secondary school. I want them to do well, which means a job in the city, and German is the language of those they must compete with there.
>
> *Mayor of Burgenland, a Croatian village in contemporary Austria*

Europeans and Others

Europeans have for many centuries moved freely over each other's lands, contributing their distinctive patterns to the cultures of different countries. Czech and Polish cultures are unimaginable without considering the contribution of Germans, French culture is unthinkable without the contributions of Italians and Spaniards, and one cannot make sense of English culture without understanding the contributions of Huguenots, the Irish, Scots, and the Welsh. Even in the era of nationalism, in which states aspired to homogeneous national cultures, migrant European groups continued to settle in different European countries and to affect the cultures of those countries. In France, for instance, to the earlier English and Germans were added, in the nineteenth century, Italians, Spaniards, and Belgians, followed in the twentieth by Portuguese, Poles, Romanians, Russians, and other east Europeans. Each made their respective contributions to "the French melting-pot"; each modified to a lesser or greater extent "French national identity" (Noiriel 1996).

These were all, of course, Europeans, mostly of Christian belief or Christian origin. There have been other groups who have also made major contributions to the cultures of European lands but whose presence has been viewed more problematically. The Jews are one example; Muslims are another. It would be hard to decide who has had the greater impact or influence—the Jews, especially in central and eastern Europe, the Muslims in the form of Arabs in Spain, Mongols in Russia, or Turks in the Balkans. All we can say is that European culture would have been immeasurably different, not to say infinitely poorer, without them. Put another way, Europe has always been a transnational space.

The Jewish Question in Europe was infamously solved in Nazi Germany. Muslims present a different problem. Seen in a series of historic encounters with Europeans, from the battle of Kulikovo in 1380 to the lifting of the siege of Vienna in 1683, Muslims have returned in more modest guise as "guest-workers" and as immigrants from the former colonies of European empires. As Turks, Tunisians, Moroccans, Algerians, Surinamese, Indonesians, Indians, and Pakistanis, they have taken up employment, residence, and in many cases citizenship among Europeans.

Muslims are not the only new workers and new citizens of Europe, of course. There are Sikhs and Hindus in Britain and peoples of various faiths from sub-Saharan Africa and Southeast Asia in several European countries, such as Angolans in Portugal and Vietnamese in France. But Sikhs and Hindus are largely restricted to Britain, and many of the other non-European groups are Christian. This does not necessarily lessen the problems they face in their new homes; and religion is only one dimension of ethnicity. But it does serve, at least initially, to set off all these other groups from Muslims. Muslims are not only the most numerous of the new immigrant populations; in their culture they seem the most distinctive and—to many in the host cultures at least—the most difficult to absorb into the host society. Muslims have become the new "other" of Europe, replacing the Jews of an earlier era and the Communists of more recent times

(Bjørgo 1997, 67). The Salman Rushdie affair, expressing itself in demonstrations by outraged Muslims and resentment and bewilderment on the part of their fellow citizens in Britain, neatly encapsulates the predicament, as does, on a lesser scale, *l'affaire des foulards*—the headscarves affair—in France (Thomas 1998; Favell 2001b, 174–83; Siedentop 2000, 207–9).

An earlier set of encounters between Muslims and Europeans permanently altered the contours of European culture and set European civilization on a new course. Will the same thing happen again? Taking Muslims as the touchstone, will the new wave of migrants and settlers—the peoples who have arrived mostly since the Second World War—be the harbinger of a new identity or identities for Europeans? Or will it instead confirm the stubbornness and persistence of old identities? Is the model to be ***assimilation, integration,*** or ***multiculturalism***—and what do these terms signify? What implications do the newcomers have for the historic role of the nation-state as the crucible of the national culture? Do they find their place within it, modifying as of old the character of the culture without seeking to replace or undermine it? Or do they rather look beyond (or beneath) the nation-state, finding in the currents of globalization the opportunities and resources to construct cosmopolitan or transnational identities? Whatever the choices and constraints, one thing is clear. The adaptation of the newcomers to their European environment will have—indeed already has had—profound consequences for the longer-established inhabitants, forcing them to rethink their own identities and perhaps invent new ones. In doing so, both they and the newer arrivals may come to see that they are in effect repeating an old story, one in which Europe and Europeans were constantly modified by the interaction of groups across borders and civilizations.

■ European Culture and European Identity

The newcomers have to adapt to Europe, even if they wish to change it in various ways. What kind of entity is this Europe? Is it unitary, marked by a fundamental character that unites the majority of its members? Is it plural, a thing of differences, diversity, and even divisiveness? In either case, is there something we can call a European identity, over and above common cultural characteristics? It is obvious that the answer to these questions will help us to assess the possible range of responses open to the new non-European groups. It will also suggest some possible responses by Europeans themselves, in the face of challenges posed by the new groups.

Whether or not there is, strictly speaking, a European identity, there is no doubt that there is something called European culture. Undoubtedly, too, the main basis of that is religion—specifically, Christianity (Dawson 1960; Eliot 1962). It is this that permits us (if we wish) to include Russia in European civilization. Russia for many centuries guarded Europe's northern flank against invaders from the east, just as Austria guarded its southern flank. While the Habsburgs, as Holy Roman emperors, saw themselves as continuing the unifying imperial mission of Rome, Moscow too claimed for itself the title of "the Third Rome," following the fall of Constantinople to the Turks. It was as the champions of Christianity against the pagans that the Tsars formulated a destiny for Russia's expanding empire (see, e.g., Averintsev 1991).

The importance of religion as a unifying factor is underlined by a consideration of how Europe was formed, what gave it its shape as a fragment of the Eurasian landmass. Europe was made by the encounter with and resistance to other religions—specifically, of course, the Muslim religion. It was largely in response to the Muslim threat—from Mongols and Tartars in the north, Arabs and Turks in the south—that Europe drew together (see Neumann and Welsh 1991; Yapp 1992). The nearest thing to all-European enterprises were the joint actions against Muslims, from the Crusades of the Middle Ages to the defense of the Habsburg empire against the Turks in the late seventeenth century.

Religion, even of the same kind, of course divides as much as it unites—contemporary Northern Ireland is there to make the point still, even if we do not wish to go back to the seventeenth-century wars of religion between Protestants and Catholics or the "Great Schism" between eastern and western Christianity that took place in the eleventh century. And there are other formative European experiences that, in the unevenness of their effects, create barriers between some Europeans even while linking others. The Roman Empire, so crucial in the formation of European law and administration, did not include large sections of central, northern and eastern Europe (while, in its eastern wing, it included much of the Near and Middle East that later fell under Muslim domination). Orthodox Europe did not, at least until quite late, share in the cultural experiences of the Renaissance, the Reformation, the Scientific Revolution, and the Enlightenment. Industrialization divided the continent into an advanced, "developed" western half and a backward, peasant, eastern half. Coming closer to our own times, Communism and the cold war further divided Europe into East and West, a divide fortified until recently by the European Community and the European Union as a western European club. These differences have at various times created a number of major east-west "fault lines" in Europe, particularly in the elaboration of a concept of Western civilization from which eastern Europe was excluded (Davies 1996, 7–31).

So there is plenty of ammunition available to those who wish to deny that there is anything like a common European identity, who say that Europe is nothing but a congeries of ethnic or national identities. At best they might be prepared to see Europe as "a 'family of cultures' made up of a syndrome of partially shared historical traditions and cultural heritages" (Smith 1992, 70). At worst they see the whole idea of Europe as a myth or an ideology—perhaps a counterrevolutionary ideology, elaborated in different ways and at different times by Western elites in the consolidation of their own power. In other words, "Europe" is a sham (see, e.g., Nederveen Pieterse 1991; Shore and Black 1994; Delanty 1995).

No one would wish to deny the differences and divisions that mark European civilization—two world wars in the twentieth century are sufficient testimony to this fact. But first, conflict, as is well known, tends to occur most fiercely between otherwise like-minded people—between brothers, families, clans, and nations sharing a common past (an effect that Freud attributed to "the narcissism of small differences"). And second, variegated and divergent patterns might be said to be the hallmark of any culture—any culture, that is, above the most primitive. For François Guizot, in his celebrated *History of Civilization in Europe,* European civilization marked itself off from all past civilizations precisely by its principle of diversity, which, paradoxically, also gave it its unity. "Modern Europe presents us with examples of all systems, of all experiments of social organization; pure or mixed monarchies, theocracies, republics, more or less aristocratic, have thus thrived simultaneously, one beside the other; and, notwithstanding their diversity, they have all a certain resemblance, a certain family likeness, which it is impossible to mistake" ([1828] 1997, 30).

So conflict and diversity are no barriers to considering the idea of a common European culture. When Edmund Burke argued that "no European can be a complete exile in any part of Europe" (cited in Davies 1996, 8), he was the last person to look for uniformity of character or harmony of culture. There can be an idea of Europe that accepts difference, even profound divisions. What presumably holds Europe together, as an ideal or ideology at least, is some notion of a common framework of acceptance, some sense of mutual understanding and recognition based on certain historical and cultural legacies. Just like national identities, a European identity can be as much a matter of habitual norms and practices as it is of some deeper cultural unity (Weale 1995, 218–19).

Such considerations of unity and diversity, distance and nearness, bring us back to Islam, and to religion. For it must be evident that a major reason for Europe's fear of and hostility toward Muslim peoples is the fear and anxiety that comes from closeness. Islam shares with Christianity a common inheritance in Judaism. Muslims have contributed to the formation of Europe not just as Europe's "other" but directly, as philosophers, scientists, and scholars. What would the European Middle Ages have been intellectually without the works of Avicenna, Averroes, Ibn Khaldun, and the other Arab thinkers who transmitted and developed the science and culture of Greece and Rome? What is Balkan civilization but a synthesis of Slavic traditions and Ottoman culture? What, at a more homely level, would European cuisine be like without the oranges, lemons, spinach, asparagus, aubergines, artichokes, and pasta introduced by the Muslims in Spain (Davies 1996, 256)?

The question of the impact of Muslim communities on European societies today must be seen within the context of this ambivalent history. Hostile encounters with Muslims have to a good extent defined Europe; but Islam has also been a part, perhaps a constituent part, of European civilization. Islam is "not-Europe," but it is also, to an extent, in Europe. The contradictions and confusions of this position affect both parties to the encounter.

There are further antinomies to consider. Europe has been transnational, most readily seen in the multinational empires—Habsburg, Hohenzollern, Romanov, Ottoman—that occupied much of its space until the nineteenth century (Lieven 2001). The space of empire created permeable and fluid boundaries that allowed for a considerable mingling of peoples, including those of non-European origin. But Europe since the late eighteenth century has also been national to an increasing extent. It has come to be defined by the nation-state, an entity that at least in part puts the stress on cultural or ethnic uniformity (Gellner 1983).

This presents the new peoples of Europe with both a choice and a dilemma. Do they, in coming to terms with their host cultures, appeal to the transnational traditions of Europe, suitably adapted to a postimperial world? Or do they try to negotiate the thickets of national politics and culture, hoping to find space for their particular identities — however conceived — within the practices of the nation-state? This is not necessarily an either-or choice. Nation-states, some of them at least, incorporate transnational understandings and practices. In current circumstances the interests of newcomers may well best be served by a strategy of working within the traditions of particular nation-states, attempting to modify them where necessary in the direction of greater tolerance and inclusiveness.

■ National, Subnational, and Transnational Identities

Europeans have liked to be able to say, "I am European, but I am also British, and perhaps also English, and perhaps even Cornish." Europe-wide identities have been accompanied and, in the last two hundred years or so, often overshadowed by national and local identities. For some, such as Anthony Smith (1995a) and Montserrat Guíbernau (1996), national identities remain preeminent, even in this era of globalization and attempts at supranational citizenship, such as with the European Union. The arrival of non-European ethnic groups only adds to the mélange of "distinctive *ethnies* and counter-cultures, of indigestible minorities, immigrants, aliens and social outcasts."

> The sheer number of these minorities and the vitality of these divided *ethnies* and their unique cultures has meant that "Europe" itself, a geographical expression of problematical utility, has looked pale and shifting beside the entrenched cultures and heritages that make up its rich mosaic. Compared with the vibrancy and tangibility of French, Scots, Catalan, Polish or Greek cultures and ethnic traditions, a "European identity" has seemed vacuous and nondescript, a rather lifeless summation of all the peoples and cultures on the continent, adding little to what already exists. (Smith 1995a, 131)

Even those, such as Jürgen Habermas, who accept the idea of a common European civilization and urge the development of a "European constitutional patriotism" as the answer to the obsolescence of the nation-state, are inclined to stress that this is more a project for the future than something already achieved and inherited. "Our task," says Habermas, "is less to reassure ourselves of our common origins in the European Middle Ages than to develop a new political self-confidence commensurate with the role of Europe in the world of the twenty-first century" (1992a, 12). Europe, in other words, does not exist, at least as a usable identity; it has to be created. In the process it must rid itself of the heritage of cultural exclusiveness and, by a process of political dialogue among its constituent groups, aim at the construction of a political entity whose hallmark is citizenship, not ascriptive membership (Robins and Aksoy 1995).

But there is a more radical posture, which has found powerful exponents. This too accepts that the nation-state, as historically formed, is outmoded. But for its advocates, the answer lies less in the creation of a pan-European identity than in going both beyond and beneath Europe. National identities are indeed losing their exclusiveness — if they ever really had such a character — and are now increasingly accompanied and perhaps superseded by other forms of identity: subnational, regional, supranational, and transnational. The usual position is to argue for a combination of some or all of these. Writing of Catalonia, for instance — a "nation without a state" — Manuel Castells has envisaged a far-reaching pattern of interlocking and overlapping identities, based on subnational, national, and transnational institutions.

> Declaring *Catalunya* at the same time European, Mediterranean, and Hispanic, Catalan nationalists, while rejecting separation from Spain, search for a new kind of state. It would be a state of variable geometry, bringing together respect for the historically inherited Spanish state with the growing autonomy of Catalan institutions in conducting public affairs, and the integration of both Spain and *Catalunya* in a broader entity, Europe, that translates not only into the European Union, but into various networks of regional and municipal governments, as well as of civic associations, that multiply horizontal relationships throughout Europe under the tenuous shell of modern nation-states. (Castells 1997, 50; see also Keane 1995, 198–205)

Such a conception, suggests Castells, though historically rooted in the Catalan case, points the way generally toward the pattern of identity in the information age. It fits a world in which the nation-state has to find its place in a global economy and a variable and interpenetrating set of cultures. No longer can the nation-state claim a monopoly over the control of welfare and culture, nor by the same token can it demand the total and unconditional allegiance of its citizens. There are other citizenship roles to play, other arenas, which provide the opportunity and resources with which to play them, and other institutions that have a legitimate claim on our loyalty.

Castells does not here incorporate the new non-European immigrant groups into his model, though it would not be difficult to do so. Indeed, Yngve Lithman suggests something similar in his account of how immigrant communities are reshaping the life of European cities. The model of national societies "assimilating" or "integrating" their immigrant minorities is, he argues, obsolete. The "Europeanisation of the nation state," together with the globalization of the economy and "a move from the notion of nation to more individualised . . . notions of personhood," are breaking down the ability of the nation-state to offer many of the traditional employment and welfare services that both made it necessary for immigrant groups to seek integration within the nation-state and at the same time allowed the state to claim their full allegiance. "It is no longer unproblematic to see state and state policies more or less harmonise with other institutions in society to create vehicles for ambitious integration (however defined) for immigrants" (Lithman 1997, 76; see also Wacquant 1996). Especially given the difficulties that they may face in the national community, immigrant communities must see themselves as free to determine their own identities, within but also beyond the nation-state. In doing so they may offer a new vitality and cultural richness to their host communities, by providing "a kind of self-help cultural 'smorgasbord,' where everyone is welcome to choose his or her favorite dishes, and where ethnic prejudices and stereotyping have been whittled away in the closeness of everyday interaction" (Lithman 1997, 78).

An even more ambitious strategy and outcome, breaking out of the European framework altogether, is suggested by Yasemin Soysal's account of what she sees as the emergence of "postnational" and "transnational" identities among immigrant groups in Europe. She too regards adaptation and integration, and especially assimilation, as outmoded strategies, in any case doomed to failure. Accommodation with nation-states still remains the goal of most immigrant groups; but increasingly such groups are drawing the resources for group identity from the universalist discourse of human rights and their own international organizations and movements. Soysal notes "an underlying dialectic of the postwar global system: While nation-states and their boundaries are reified through assertions of border controls and appeals to nationhood, a new mode of membership, anchored in the universalistic rights of personhood, transgresses the national order of things" (1994, 159).

Turks in Germany, for instance, especially second-generation Turks, will be neither Turkish nor German. They will not model themselves on the culture either of the sending or the receiving society. They will—as they already have done—follow a third path, building their identities out of the discrete elements of the host society, their home society, and the postnational possibilities made available to them in the language and institutions of universal human rights. "Multiculturalism, the right to be different and to foster one's own culture, is elementally asserted as the natural and inalienable right of all individuals." Old particularities are now clothed in the new human rights discourse. "Mother Tongue Is Human Right" was the slogan of the Initiative of Turkish Parents and Teachers in Stuttgart, in pressing their claims for Turkish-language instruction in German schools (Soysal 1994, 154–55; see also Faist 1998).

The paths marked out by Castells, Lithman, and Soysal are in no way unreal or implausible. They dovetail well into much of the recent thinking on migrant communities and multiculturalism. John Rex, for example, says that the migrant ethnic community, based on the extended family and stretched across several nations, is "today at least as important as membership of a nation." He instances the Punjabis in Britain, Canada, and the United States. The members of such a community, moreover, "belong to different social groups and different cultural systems simultaneously and are used to the fact of multiple identities" (1995, 29, 32; see also Rex 1994). Multiculturalists, whether or not of a postmodern persuasion, are also much given to talk of multiple or "nested" identities. Not just migrants, they say, but all individuals in contemporary conditions live at the intersection of several cultures, from which they are free to select in composing their identities—and in recomposing them at a later date. "Not only societies, but people are multicultural," declares Amy Guttman (1993, 183). We all embody several overlapping identities; it is bad faith to attach ourselves too exclusively to any one of them, such as national identity. In this sense migrants and migrant communities are the outriders of the emerging postnational society—they are, no less, "a metaphor for all humanity" (Rushdie 1992, 394; see also Hall 1992, 310–14; Appadurai 1996, 139–77; Wieviorka 1998).

Part of the attraction of these positions, no doubt, is that they avoid the essentialism of earlier writing on multiculturalism: the assumption that ethnic cultures were primordial and fixed, and that multiculturalism consisted in a patchwork quilt of discrete cultures, each preserving its original form and more or less stable identity (see Çağlar 1997). By contrast we are invited to consider a future of "hyphenation," "hybridity," "syncretization," "creolization," and the creative inventions of "diaspora cultures" (see, e.g., Gilroy 1987; Nederveen Pieterse 1994; Appadurai 1996; Faist 1998). Immigrant communities, in an eclectic mixing of the resources at hand, will transform not only themselves but also the societies they inhabit—for the better. The goal is not "assimilation"—"something immigrants or minorities must do or have done to them"—but rather "integration" as "an interactive, two-way process; both parties are an active ingredient and so something new is created." In the "plural state," as opposed to the older conception of the liberal state, "multiculturalism means reforming national identity and citizenship" (Modood 1997, 24).

We have already considered that European culture is more porous and permeable than often thought. It is fissured through and through. It already has a generous leavening of non-European, specifically Muslim, culture, as a result of past encounters. It is not set in a rigid mold. Does this not create a space for the gradual transformation of European societies? Might not the new immigrant communities, in making and remaking themselves in their new environments, be the seedbed of social and cultural innovation in the wider societies they inhabit? Might they not indeed be the building blocks of a new Europe itself, undermining "from the bottom upwards" the nation-state just as the European Union is attempting to do so "from the top down"? Will the new Europe be one of "ethnic heterogeneity inserted into a multicultural suprastate" (Modood 1997, 1; see also Werbner 1997, 263; Zincone 1997, 132)?

■ The Nation-State, Citizenship, and Immigrant Communities

The attractions of these and similar formulations should not blind us to some obvious stumbling blocks. Most of these center on the persisting strength and salience of the nation-state, at least as perceived by many of the nation-states' inhabitants. Soysal herself admits that "the nation-state is still the repository of educational, welfare, and public health functions and the regulator of social distribution" (1994, 157). It is also still the only legitimate political form recognized by the international system. This means that in practice many supposedly universalistic human rights—not just the right to vote but the right to work, to education, to health, to housing—have to be claimed and if necessary fought for within the framework of the nation-state (Morris 1997, 197–98; Joppke 1999; Koopmans and Statham 1999). Most immigrant communities seem to recognize this, even if they also make use when possible of international agencies — such as the International Court of Justice — in pursuing such claims. It may partly explain why some of them, for instance those in Britain, are very ambivalent about the pull of Europe, seeing in further European integration the possible loss of hard-won gains as much as the opportunities for new benefits (Favell 2001b, 210–11).

This points, in turn, to the continuing importance of citizenship. Writers in the transnational or postnational vein have argued that what matters most these days is not citizenship but residence. Permanent residence, they say, coupled with the appeal to universal rights, has largely replaced citizenship as the basis for the claiming of social rights. This suggests an ingenious reversal of T. H. Marshall's famous historical trajectory whereby European populations moved from the winning of civil and political rights to the achievement, mainly in the twentieth century, of social rights. Now, it is argued, largely through the pressure of international human rights organizations, social and economic rights have come to be conferred irrespective of formal citizenship. "The transformation of 'national' rights into more universalistic entitlements that include noncitizens," says Soysal, "undermines the categorical dichotomies patterned after the national citizenship model" (1994, 135). David Jacobson (1996, 8–9) argues that "as rights have come to be predicated on residency, not citizen status, the distinction between 'citizen' and 'alien' has eroded. . . . Citizenship . . . has been devalued in the host countries; aliens resident in the United States and in Western European countries have not felt any compelling need to naturalize even when it is possible" (see also Favell 1997, 189; 2001b, 242–50; Bauböck 1998). The way forward for immigrants is not to press for political citizenship in the states in which they reside but to urge greater recognition of their universal human rights as persons.

This could be a dangerous illusion. In pursuing transnational or universal rather than national citizenship, immigrants may condemn themselves to passivity and to second-class citizenship within nation-states that still hold most of the levers of power and control over the most important aspects of their lives. As even the advocates of transnational citizenship admit, "the state itself is the critical mechanism in advancing human rights" (Jacobson 1996, 11). It is left to the state to translate into policies and bureaucratic practice the injunctions and declarations of international human rights institutions. Here is the rub. As Lydia Morris shows, despite the efforts of the European Union, national governments in

Europe have continued to act restrictively over such matters as immigration policy, the rights of migrants, citizenship in the European Union, and human rights. Several states, under the pressure of labor and other constituencies, have found ways of restricting access to welfare systems and labor markets to those who, though resident, are not citizens (Morris 1997, 205–6). Noncitizens, of course, do not make up voting constituencies; national political parties do not therefore have to pay much attention to them. There are votes to be gained in standing against immigrants; virtually none in standing for them.

Immigrants in Europe have only to look across the water, to the United States, for a clear warning of the danger of espousing residence rather than citizenship. Here for many years legal or permanent residents felt they had equal rights with citizens and that naturalization not only was not necessary but incurred unwanted burdens (Goodwin-White 1998, 420). Legal residents experienced a rude awakening in 1997, when by act of the U.S. Congress they found themselves deprived of most rights to social security benefits. Not surprisingly, many permanent residents have since been rushing to acquire citizenship (a trend started in the midnineties in anticipation of this and similar measures—see Goodwin-White 1998, 421). In these days even citizenship is no guarantee against the infringement or deprivation of rights. But it offers at least some sort of protection, if only against deportation at the arbitrary whim of immigration officials (see "Immigrant Lockup" 1998 for the increasing practice of this in the United States, in relation to both legal and illegal residents). Perhaps the French are right, after all, despite the criticisms heaped upon them by the transnational theorists, to insist upon the overriding principle of citizenship for all (for the French position, see Silverman 1992; Hoffman, 1993; Favell 2001b, 150–99).

If immigrant communities are compelled to face up to the realities posed by the power of the nation-state, a further difficulty for views of postnational or transnational citizenship comes from the rooted nationalism of most Europeans themselves. There are significant national differences here—Italians and Germans are more pro-Europe than the British and French, for instance (Borri 1994). Yet, polls have repeatedly found that, despite the efforts of the European Union and a vast increase in social interchange and media integration, the nation-state remains the principal focus of popular identity and the main basis of legitimacy. As William Wallace summarizes the evidence: "A certain diffusion of loyalties, a certain expansion of horizons from the national to the European (and the global), are evident both among elites and — more faintly — among mass publics. But challenges to the legitimacy of national institutions and elites have come largely from within existing states: leading to fragmentation, not integration. Throughout Western Europe the national community remains the broadest focus for political life and group identity" (1994, 55–56; see also Schlesinger 1992a; Mann 1993).

Manuel Castells, too, has pointed to the fact that "since the [European] integration process has coincided with stagnation of living standards, rising unemployment, and greater social inequality in the 1990s, significant sections of the European population tend to affirm their nations against their states, seen as captives of European supranationality" (1998, 326). Political elites in the European states tend to be pro-European, but their mass publics are much less so, when they are not actually hostile to the European project. Johan Galtung some time ago noted the phenomenon that just as the elites of postindustrial societies were "growing out" of the nation-state, the mass of the population was, perhaps for the first time in their history, coming to identify fully with the nation-state and to defend it energetically against threats from both within and without (1969, 19–20). A similar view has recently been expressed, rather more caustically, by Christopher Lasch (1995). The irony is that just as processes of globalization and internationalization are most profoundly affecting the lives of all groups and classes of society, some of those groups, in a defensive reaction, are feeling the need to insist more strongly than ever before on their national identity. "Nationalism, not federalism, is the concomitant development of European integration" (Castells 1998, 327).

The persistence of nationalism and the nation-state must clearly affect the attitudes of newcomers and native Europeans toward each other — to return to our original concern. Muslims, and other ethnic minorities, see native Europeans acting in the name of an often aggressively restated definition of themselves as English, French, German, and so forth. These definitions contain ethnic terms that generally exclude non-Europeans. Primarily they may be stated in terms of color, but they also usually include cultural elements that defy, and are meant to defy, attempts by non-Europeans to adopt them or adapt to them. Thus a former British Conservative cabinet minister, Norman (now Lord) Tebbit, formulated what has come to be known as the Tebbit Test (McGuigan 1996, 138): do "black Britons" support the "English [*sic*]" cricket team when it plays against India, Pakistan, or the West Indies? If not, they cannot claim to be British (this ignores the number of black Britons already on the English team, just as a similar test involving football in France would ignore the number of North African French in the national team). French nationalists have denounced the wearing of scarves by Muslim schoolgirls on the grounds that this goes against French secularism — thus showing, once again, that secularism is not a neutral category but carries a distinct cultural charge (Silverman 1992, 111–18).

The obvious retort to all this on the part of non-European minorities is to refuse the labels imposed on them — as incomplete or inferior British or French — and to seek identities that draw upon both past cultural origins and the new resources available to them in the international arena. That, as we have seen, is the strategy advocated by Soysal and others. The problems in this case may be equally grave. There seems to be an inescapable dilemma here. One asserts an identity that involves, for instance, retaining one's language and religion as central elements. But one lives in a national society that defines itself predominantly in other terms and at the same time controls the points of entry to the principal sources of economic opportunity and political participation. How does one compete on equal terms with those nationals who are adept at deploying the master symbols and cultural styles of the dominant society? Are minorities not condemning themselves, in practice if not principle, to second-class or even ghetto status — as members in but not of society? It is this problem, incidentally, that drives a coach and horses through the hallowed distinction between civic and ethnic belonging: How does one truly participate in the civic realm without possessing the requisite cultural resources to perform there successfully, and on equal terms with all citizens? What does it mean to press for a "French Islamic civilisation" (Diop 1997, 125) or a "British Islam" (Lewis 1994)?

There is a further consideration. Many have pointed to the way in which nation-states, as they lose control over large areas of economic and cultural life, have responded by intensifying their hold over those areas that they remain more or less in command of. These include power over the granting of citizenship and nationality, together with associated power over the issuing of passports, visas, residence and work permits, and so forth. States now display themselves — there seems no better word for it — at borders and frontiers, controlling movement in and out, demonstrating by elaborate policing and surveillance the continued relevance of national boundaries (Mehta 1999, 151). Earlier-established Europeans, at least west Europeans, may feel that things have become much better for them in this respect, especially those who find themselves in the countries of the Schengen agreement. But for new Europeans and those still seeking citizenship, especially if they are of non-European stock, the situation is very different and can be the cause of great hardship and frustration.

This is as true of European citizenship—more precisely, European Union citizenship—as it is of national citizenship. The establishment of European citizenship was one of the most vaunted achievements of the 1992 Treaty of Maastricht. At a stroke more than 90 percent of European Union residents gained European citizenship, in addition to their national citizenship. But as Elspeth Guild has shown, of all the provisions of the treaty, those relating to nationality and citizenship are the vaguest and most open to particular interpretations (or evasions) by national governments. For instance, no right to a European, any more than to a national, passport exists: "issue and withdrawal conditions apparently remain a national prerogative" (Guild 1996, 48). European citizenship is a derivative status, dependent upon citizenship in one of the member states of the European Union. Since national governments keep control of the power to grant or withhold national citizenship, they effectively also keep control over the power to grant or withhold European citizenship, thus rendering the whole concept rather vacuous (Morris 1997, 198; Shore and Black 1994, 282; Brewin 1997, 232–34).

That this can matter even to those who seem in most ways highly privileged members of European societies has been brought home with especial force to one particularly keen-eyed citizen. Michael Ignatieff was, by his own account, proud to call himself a cosmopolitan in an increasingly unified and multicultural world. Anything else seemed a rejection of the Enlightenment philosophy by which he lives. But his travels in the troubled places of the earth have made him realize the limits of this lofty position. Bosnia in particular taught him a chastening lesson in realism.

> What has happened in Bosnia must give pause to anyone who believes in the virtues of cosmopolitanism. It is only too apparent that cosmopolitanism is the privilege of those who can take a secure nation state for granted. Though we have passed into a post-imperial age, we are not in a post-nationalist age, and I cannot see how we will ever do so. The cosmopolitan order of the great cities—London, Los Angeles, New York, Paris—depends critically on the rule-enforcing capacities of the nation state. . . .

> In this sense . . . cosmopolitans like myself are not beyond the nation; and a cosmopolitan, post-nationalist spirit will always depend, in the end, on the capacity of nation states to provide security and civility for their citizens. (Ignatieff 1994, 9)

Ignatieff's discovery that, when he came to the security checkpoints and other dangerous obstacles in his travels, it was his Canadian passport as much as the attendant television cameras that gave him the freedom to pass and move about the country, is an experience shared at different levels by many who have never put themselves in such threatening situations. It illustrates vividly the continuing power of national citizenship and the difficulties faced by those who in various ways may seek to go around or beyond the nation-state. Postnational identities and transnational citizenship are undoubtedly attractive goals; they can be achieved in part not just by secure and wealthy cosmopolitans but, no doubt, by immigrant communities seeking to protect and extend their particular identities. But a definite element of wish-fulfillment seems to have entered into the thinking of those who, in current circumstances, proclaim them to be the best or only avenues of advancement.

Europe has always been a promise, an ideal, even an ideology, as much as it has been an achieved reality. Its legacies have been ambivalent. They have included a tradition of assimilation and cultural pluralism that militates against any strict delimitation of boundaries, whether ethnic or territorial. Large, loose political entities have encouraged cultural mixing both between long-resident European groups and between these and other groups that originated beyond Europe's always shifting borders.

Europe also gave rise to the nation-state. This has in some cases meant a hardening of boundaries and attempts at cultural and ethnic homogeneity. But in several cases—those in which understanding of nationhood more approximates to the civic—it has also meant the incorporation in the state's own laws and practices of beliefs and values that confer considerable power on groups to shape their own identities and to establish varying relationships with the dominant culture. These include precisely the notions of human and civic rights that transnational theorists wish to derive from international organizations. As the European Union strengthens its institutions and generates greater commitment from among the peoples of Europe, no doubt minority groups along with others will increasingly look beyond the nation-state—and perhaps even beyond Europe—for the fulfillment of their aspirations (see, e.g., Schierup 1995). But they would be wise to wait for supranational institutions to realize their promise before they abandon the strategy of seeking to make the nation-state live up to its own ideals.

NOTE

I should like to thank Mabel Berezin for her many helpful suggestions in the revision of this chapter.

TWO

Political Community, Identity, and Cosmopolitan Place

J. NICHOLAS ENTRIKIN

The moral significance of place becomes evident when place is conceived not as location in space, but instead as related to an individual subject. Through the subject, place draws together the object realms of nature, society, and culture. This relational concept of place contributes not only to the understanding of self and identity, but also to the constitution of collective identity through territorially based communities. In sociology the linkage of place and community has often been framed in terms of human ecology, but one could generalize from Fernand Braudel's admonition to French sociologists when he stated that the use of the term *ecology* is simply "a way of not saying geography" (1980, 51). Both geographers and ecologists draw together object realms that theories based in other disciplines tend to separate. The geographer's place concept, however, extends beyond the naturalism of the ecologist to include an explicit relation to a subject (Sack 1997; Tuan 1977; Entrikin 1991).

Through place, the subject is related to milieu or environment. Just as the individual subject is embodied, it is also emplaced (Casey 1997). Place, understood as the necessary context for human actions rather than their mere setting, bears a reciprocal, mutually constitutive relation to the self (Sack 1997). It is this necessary relation to self that extends the meaning of place beyond that of objects located together in space and into the realms of the moral and the aesthetic. This relational place concept is thus implicated in the formation of solidaristic ties and the construction of political community.

My goal in this chapter is to suggest how one might link the geographical and philosophical discourse on place with that concerning democratic political community. The first step is to discuss the idea of democratic place-making and to note the sometimes explicit but more often implicit role that place currently plays in discussions of political community. For the most part, such discussions link place with ascriptive and primordial qualities and thus with particularism. The geography of political community will then be illustrated in a brief discussion of the moral geographies that are part of the debate surrounding the formation of the European Union. In the conclusion I offer an outline for a more cosmopolitan conception of place that moves beyond the local and particular toward the universal.

■ Good Places

Benjamin Barber (1998, 3) uses place as a metaphor for civil society in his characterization of the challenges facing modern democracies:

> Our world, on the threshold of the millennium, grows crowded: too many people, too much anarchy, too many wars, too much dependency. Plagued by the effects of this crowding . . . we look, often in vain, for hospitable spaces to live in, for common ground where we can arbitrate our differences or survive them with civility, for places where we can govern ourselves in common without surrendering our plural natures to the singular addictions of commerce and consumerism. A place for us, that is all we seek. A place that allows full expression to the "you" and "me," the "we" of our commonality, a place where that abstract "we" discloses the traces that lead back to you and me.

Barber, like other political theorists, implies that good places are constructed through civil society, or, stated another way, a good and just society produces good and just places. But the complexity of place as both a material and a mental reality resists such easy classification, as much of the contemporary literature on environmentalism suggests. The construction of good places in democratic societies involves a relation to the natural world just as it involves the creation and maintenance of human institutions. More importantly for this discussion, to construct new places means that existing places are destroyed or changed, and to maintain places that allow certain practices means that other practices are excluded (Sack 1999). This interplay of presences and absences shapes the experience of place and influences human practices. To view place in this way helps one to see that the actual places of civil society do not simply follow from political and social practices but also enable and constrain such practices.

The association of geographical and social ideals has been a common theme of cultural myth and religious belief and has a long intellectual history in moral and political philosophy (Glacken 1967; Tuan 1989; Casey 1997). Modern thought can thus draw upon a substantial intellectual reservoir of varied and sometimes conflicting arguments about what constitutes a good place. For example, in Judeo-Christian doctrine, the idea of the good is related to the perfection of the divine and the desire of imperfect humans to approximate this unattainable standard. In classical Greek philosophy, this perfection is associated with both a sense of completeness or wholeness and the achievement of a goal or end, as in the Aristotelian sense of realizing a specific form or a "natural place" (Casey 1997; Owen 1974). These same themes are evident in early Christian thought, with the recognition that the City of Man could never achieve the goodness of the City of God (Augustine 1972). Completeness was only possible in the divine.

In modern democratic societies, this linkage of the good with some notion of perfectibility seems archaic and potentially dangerous. The twentieth century offered several examples of teleology and visions of completeness put into the service of murderous absolutisms, in which organic wholeness was a central element of totalitarianism (Arendt 1986, 437–59; Hutchings 1996, 81–84). For liberal theorists such as John Rawls (1971, 325–32), perfectionism has no positive role to play in democratic societies.

Another quality of modern thought that bears on this issue is the emphasis on process over teleology, on contingency and change rather than final ends and completed states. In the early nineteenth century, Tocqueville pointed in this direction without fully abandoning teleology when he noted the intellectual propensity in a then youthful American culture toward "a lively faith in the perfectibility of man" and a belief in "humanity as a changing scene, in which nothing is, or ought to be, permanent; . . . [Americans] admit that what appears to them today to be good, may be superseded by something better tomorrow" ([1835] 1990, 393).

A century later, this cultural quality was incorporated into a process metaphysics in the writings of John Dewey (1927, 143–44), who emphasized democracy as a way of life and as human practices rather than as a set of institutions. As a way of life it began at home and in the community, at school and in the factory. Liberal education prepared individuals to take an active and critical role in this building of place and community by instilling a cosmopolitan spirit that allowed one to transcend the here and now, to be part of a conversation with past generations and distant cultures. Thus, one gained a sense of centeredness through active participation in community life, but that participation fostered a feeling of being part of something larger (Dewey [1916] 1944).

This cosmopolitan vision and the elusive goal of democratic place-making can be framed in the language of thick and thin as used by Michael Walzer (1994) and applied to place by Robert Sack (1997). Thick places rich in cultural traditions and customs create difference, but often through erecting highly impermeable boundaries that restrict entry and access. Boundaries here could in some instances mean political borders but could also refer to social and cultural barriers that maintain distinctions between insiders and outsiders. To the view of outsiders such places remain opaque, and in the extreme instances, they are closed to all who are not members. Thin places are more permeable and more open to view, but as a result they may lose a sense of local difference and become more like other places. This latter type of place is often associated with the homogenizing qualities of globalization, while the former is linked to cultural particularism. Returning for a moment to Dewey, both the thinning of place through the growing corporateness of American society and the thickening of place through an inward-looking provincialism hostile to the goals of liberal education were threats to healthy democratic practice.

■ Place and Ethnos

The relationships of self and community to place are now most often associated with difference, particularism, and localism. The recently more favored view of these issues has made place studies a cottage industry not only in human geography but also in the other human sciences. In the United States, this interest has corresponded with a shift in cultural attitudes about place attachment. For example, Yi-Fu Tuan (1996, 6) notes that the still powerful American belief that "no one need feel locked in place," a guiding theme in the civil rights movement and the fight against racial segregation as well as in the struggle to expand opportunities for women, has been challenged by the growing acceptance of a counterideology that makes place the locus of human fulfillment to be protected and guarded from change. This latter view is prominent in both antimodernist nostalgia for traditional community and stable identities and the postmodernist valorization of situatedness, context, and difference. Each is contrasted with the centerless space of modernism in which difference is muted through homogenizing and globalizing tendencies and place becomes mere location in space.

In political theory, this theme has been most explicit among communitarians. For example, Michael Sandel (1996) asserts that the identities that connect people to particular communities and local solidarities remain an important quality of modern social life. They persist, he says, because of, rather than in spite of, globalization pressures. For Sandel the global communications systems and markets that shape our modern world and that "beckon us to a world beyond boundaries and belonging" enhance rather than diminish the need for developing the necessary local "civic resources," which are rooted in "the places and stories, memories and meanings, incidents and identities that situate us in the world and give our lives their moral particularity" (349).

> Since the days of Aristotle's polis, the republican tradition has viewed self-government as an activity rooted in a particular place, carried out by citizens loyal to that place and the way of life it embodies. Self government today, however, requires a politics that plays itself out in a multiplicity of settings, from neighborhoods to nations to the world as a whole. Such politics requires citizens who can think and act as multiply-situated selves. The civic virtue distinctive to our time is the capacity to negotiate our way among the sometimes overlapping, sometimes conflicting obligations that claim us, and to live with the tension to which multiple loyalties give rise. This capacity is difficult to sustain, for it is easier to live with a plurality between persons than within them. (350)

According to Michael Walzer (1994, 85–104), the modern self divides into parts through differing spheres of interests, identities, and ideals. The difficulties that Sandel associated with such divided selves are worth the trouble, Walzer believes, in that they reduce the risk of fundamentalism and conflict, for when "identities are multiplied, passions are divided" (82).

Sandel's statement illustrates one side of the contemporary division in ethics and political philosophy between virtue and justice. Whereas once justice was discussed as a virtue, justice and virtue are now commonly viewed as separate and possibly incompatible realms in the current tendency to associate virtue with particularism and justice with universalism (O'Neill 1996). The related geography links virtue with ways of life rooted in place and justice with a uniform space of rights and obligations.

Space and Demos

The global interconnectedness of modern life is most often cited in relation to the world economy, but it has been an equally significant factor in stimulating international discussions on human rights and global moral responsibilities. For example, the environmental movement has spawned a number of international organizations that have sought to establish a transnational standard of duties and obligations. Similarly, human rights are perceived as international in their reach. Each of these examples points toward a new international space of rights and obligations but for the moment remains dependent on the cooperation of nation-states for implementation and enforcement. This vision of a global, homogeneous space of rights and obligations has also become part of the debate about the future of citizenship.

One element in such discussions is the desire to rid citizenship of its ascriptive and particularistic qualities. Jürgen Habermas (1996, 492) sees a fundamental "conflict between the universalistic principles of constitutional democracy, on the one hand, and the particularistic claims to preserve the integrity of established forms of life, on the other." He gives universalism and inclusion precedence over the concerns of community cultural integrity, a view shared by Veit Bader (1995, 222), who asserts that "any morally defensible concept of democratic citizenship ought to start from universalism. . . . As a first step, therefore, citizenship ought to be disentangled from ascriptive criteria and identities."

The association of place with particularism and ethnos, and space with universalism and demos, reflects the combination of two quite distinct philosophies of place and space (Curry 1996, 87–90; Casey 1997). The first derives from Aristotelian philosophy, in which all things have their natural place and in which the world of places is logically independent from and prior to an abstracted conception of uniform space. The second derives from a Cartesian emphasis on the priority of space as pure extension, in which places can be carved out of preexisting space. The uncritical mixing of these two philosophies has clear consequences for arguments about both the imagined and real geographies of political communities. For example, in American jurisprudence the issue of political representation and electoral redistricting has struggled with the constant tension between two competing views: one concerning regional community representation that emphasizes place and belonging and the other concerning numeric representation that highlights the spatial logic of districts as containers of individual rational actors. The failure to separate these two philosophical logics has created a somewhat confusing case law around the issue of racial and ethnic redistricting (Forest 1995).

These two views are also evident in discussions of building political community in the European Union. The next section offers a brief synopsis of some of the myriad arguments associated with this ongoing debate in order to illustrate the implicit geographies that are part of arguments concerning identity and membership in the EU (Entrikin 1997).

■ The European Union

Both supporters and critics of the EU have been concerned with its apparent lack of a strong sense of identity and political community (e.g., Lenoble and Dewandre 1992). Analysts have noted the EU's "democratic deficit," here referring in part to the common view of its bureaucratic (or "Eurocratic") origins and its relatively weak connections to the general populace of Europe (Pogge 1997; Andersen and Eliassen 1996). The EU has sought various ways to overcome this deficit, such as the implementation of the subsidiarity principle, which involves a vertically distributed sovereignty matching functions with the appropriate spatial scale of political community, but public indifference remains a concern (Føllesdal 1998; Pogge 1997). François Furet (1995, 87) has claimed that prior to the French referendum of the early 1990s, "the European idea grew up in France virtually incognito." Some, such as Thomas Pogge (1997), see the creation of the EU as an opportunity to develop political and social institutions democratically. Others, such as Claus Offe (1998, 116), argue that such a goal is in principle impossible for any democracy because "democracies are neither self-founding nor self-enforcing."

A useful general schema for framing this discussion is the continuum formed by the two current poles of debate on political community: liberal individualism and communitarianism. The first position is characterized by an emphasis on rational, autonomous subjects who through self-reflection are able to distance themselves from the world of social relations, and the second sees the self as shaped and given identity through social attachments and sentiments of belonging. At the extreme individualist end of this continuum is a market model of political community that draws upon neoliberal economic theory and emphasizes one aspect of the geography of Europe — the space economy. Concern with the geometries of location, the efficiencies of scale, barriers to movement, and the friction of distance becomes paramount. These kinds of concerns are prominent in publications of the European Commission (e.g., European Commission 1994) that portray the advantages of the Union in terms of an expansion of choices for the rational consumer, from the cornucopia of consumption associated with an expanded market to the greater freedom of choice about where to work and where to live. From this perspective, Europe becomes a land of the free flow of people and goods, an open market with a free-floating European population. This space of Europe is less a space of diverse places than it is a space of locations offering different "opportunity bundles" to a citizenry of utility maximizers with changeable and flexible identities and thin connections to place and regional culture.

Those who question a purely economic foundation for the EU criticize the thinness of membership in such a broadly and materialistically conceived European political community. One critic, J. G. A. Pocock (1997, 29), offers a historical view of the ideal of a unified Europe rooted in the desire for political stability and peace in that region. He contrasts the Enlightenment ideals of Europe as a partnership of cooperating states with the contemporary idea of a submergence of the state into what he refers to as a "postmodern arrangement." It is his concern that as national and civic attachments are made secondary to the forces of a global market, individuals become more important in their role as consumers than they are in their role as citizens.

Advocates of a civic as opposed to a market model have applied their arguments across a range of scales. Some extol the democratic possibilities of a world community; others, like Pocock, continue to support the sovereignty of the nation-state; and still others favor a denationalized public space of disparate communities. For example, Habermas suggests a civic model applied on the world scale. Solidarity, he argues, must be distinguished from civil society in a cosmopolitan space (1996, 514). The democratic right of self-determination allows one to protect one's political culture but not a particular cultural form of life. He offers a concept of a public space with open external borders, an element that is missing from many neoliberal, market-oriented considerations.

For Habermas, European citizenship may be seen as a transitional stage, a step on the way toward world citizenship: "Even if we still have a long way to go before fully achieving it, the cosmopolitan condition is no longer merely a mirage. State citizenship and world citizenship form a continuum whose contours, at least, are already becoming visible" (1996, 515). The civic view applied on the subnational scale is expressed by Etienne Tassin, who argues that "the idea of a European fatherland has to be replaced by that of a public space of disparate communities. A European political community will be born not so much from an 'idea' of Europe as from the idea of a public space of fellow-citizenship which is alone capable of giving meaning to a non-national political community" (1992, 189).

Such arguments have been used to support programs for political devolution in Europe, but with the latent threat that in some regions progressive intentions may inadvertently re-ignite the embers of atavistic conflict (Scott 1998, 152–57).

As one moves further toward the communitarian end of the continuum, one finds cultural pluralist models that consider ethnic, regional, and national communities to be the locus of personal and group attachments and political identity. In the most extreme form, when ethnic and national sentiments are the basis of solidarity, Europe is pictured as a composite of particularistic places and territories. This heterogeneous territory is usually associated with unassimilated cultures of various scales ranging from regions to nation-states (Smith 1991). Unlike the thin space economy of the market model with its goal of a frictionless plane of movement, or the public space of the civic model with relatively permeable borders, here one finds thick, relatively impenetrable places of cultural attachment and local, regional, or national identities. If European at all in intent, it is a model that implies a confederal future.

In this view a unified and integrated Europe becomes secondary to the goal of ethnic, regional, or national autonomy. In an optimistic if somewhat counterintuitive assessment, the particularism of this Europe of nationalities and ethnicities reduces the sources of potential conflict, in that the thick attachments and sense of belonging strengthen one's interest in protecting such communal arrangements through a mutual toleration among similarly thick communities. For example, Anthony Smith (1995b, 64) argues that "the modern history of Europe suggests that we should be prepared to accept such processes of national self-strengthening and build upon them, if we wish to secure a peace that gives reassurance to all of the peoples of Europe."

■ The Geography of Europe

It has been suggested that the different visions of political community that frame the debate about the political future of the EU have specific geographies associated with them. At the most fundamental level, the neoliberal market model and the civic model differ from the others in terms of their preference for a vocabulary of space rather than place (e.g., Cole and Cole 1993; Brunet 1989; Shaw, Nadin, and Westlake 1996). As previously noted, the market model emphasizes a space economy in which barriers to movements of goods, services, and people are reduced and the friction of distance is minimized. The civic variant places less emphasis on the economic realm and more on a homogeneous social space of the public sphere, a space that levels hierarchies, removes social barriers, and accentuates common citizenship and equally shared duties and responsibilities. In both cases, however, the geographic dimension is much the same — that of decentered space rather than centered place, of a homogeneous space of opportunity, fair access, and social justice. Both views represent a cosmopolitan vision of political community. One favors commerce as the source of a cosmopolitan spirit, with a frictionless space economy as a means toward that end. The other favors a homogeneous space of political rights and duties and access to political authority. Each opposes the lingering particularism of a Europe of nations or a Europe of ethnic groups (Smith 1995b).

The differences among these geographic conceptions become more apparent in the consideration of borders. In the market model, the internal borders of Europe disappear, but an external border is erected instead. This is the model found most often in the literature of the European Commission and in much of the public discussion of international trade issues and immigration. It is easy to understand how this image might fuel some of the concerns expressed by voters about being left outside the borders of this new Europe. In the civic model of Habermas, the internal and the external borders are transitional, eventually disappearing as European political community progresses toward world community. In the cultural pluralist model, the zones of inclusion and exclusion remain clear and marked by places of thick cultural attachments. The borders within Europe change but overall are strengthened or made increasingly impermeable, and since internal borders are conducive to difference, external borders become redundant.

The close connection of place to a subject indicates that visions of democratic political community involve not only implicit geographies but conceptions of self as well. For example, the neoliberal model posits a rational self whose actions are guided by the careful weighing of choices presented in an individualistic utilitarian calculus. In the civic model of Habermas, the self is also a rational moral actor, one who is able to step outside of self-interest, community, and culture to seek universalistic principles of social justice in constructing political community. In the radical democracy variant of the civic model expressed by Etienne Tassin and articulated by Chantal Mouffe (1992, 10), the self is constructed not as a unitary subject but as an ensemble of subject positions at the intersection of multiple, often conflicting, discourses. In sharp contrast to this anti-essentialist argument, the cultural particularist model offers a self that is determined by cultural identity and context, identities that change so slowly that they may effectively be thought of as primordial and essential when dealing with contemporary and relatively fast-developing political issues such as the rapid evolution of the EU.

■ Conclusion: Toward Cosmopolitan Place

The discussion herein has considered the concept of place in theoretical arguments about collective identity and democratic political community. Contemporary Europe was used as an illustration, but other examples could have been presented as well, such as the struggles over identity and political community in Quebec or, on a larger scale, Canada as a whole (Kymlicka 1995). In discussions of these specific cases as well as in more abstract theoretical arguments, place is most often associated with the local and the particular, and the geographic is conceived as a set of static and fixed relations rather than as processes (Agnew 1994). Stable democratic political community would appear to require places that are dynamic, malleable, open to a world beyond the local, and conducive to practices supportive of the universalistic ideals of a common humanity. If this assessment is accurate, then there seems little reason why abstract, theoretical discussion of self, community, and identity in democratic societies should leave place behind (Berdoulay and Entrikin 1998).

Recent discussions of civil society offer hints concerning what such a cosmopolitan conception of place might look like. Barber (1998, 3) points in this direction with his reference to the abstract "we" that leaves traces of a connection to a more specific "you" and "me." For Jeffrey Alexander (1997, 1998) the sense of "we-ness" that derives from a shared form of life opens up to the ideals of universal community through democratic practices. Cultures of democratic societies create a solidary sphere in which universalism need not be limited to abstract reason but may instead be presented in concrete terms. Civil society is thus understood "as a solidary sphere in which a certain kind of universalizing community comes gradually to be defined and to some degree enforced" (Alexander 1998, 7). Such a community becomes an ideal that actual communities approximate to different degrees.

Place may be understood in a similar way, in which the concrete relations linking individuals and collectives to milieus open to a more generalized and universalized sense of "here-ness," that centers on "me" and "us" but that is potentially inclusive of all humanity or, for some more extreme forms of environmentalism, all life forms. Building good places in democratic societies thus may be associated with balancing the concrete relations of attachment with egalitarian ideals. For example, the universalism embedded in democratic principles would seem to entail the elimination of borders or at least the creation of highly permeable boundaries. However, once again one faces the dilemma implied in the opposition of ethnos and demos: boundaries help create difference and communal identity, and their elimination risks creating a uniform, placeless world with weakly attached citizens. Thus, concern for both participatory democracy and toleration would seem to encourage overlapping, differentiated places of attachment with relatively permeable boundaries.

Another democratic value associated with the building of good places is the importance of making place open to view (Sack 1999). Secrecy, deception, and hiddenness coexist uncomfortably with a commitment to democratic principles, as illustrated in the continuing debates in liberal democracies over the role of covert intelligence and military activities. The opening up to view of the home, the workplace, the school, and the on-the-ground activity of the local or national state has generally been seen as a progressive step in modern democracies. Openness and access in democratic societies not only discourage abuses of power and corrupt practices but also encourage participation and an active sense of membership. However, some liberal theorists have noted that openness too has its limits, and that the need to maintain civility in complex modern societies requires that a clear line be drawn between the public and the private. For Thomas Nagel (1998), concealment of the private is not only a necessary but also a positive and constructive practice in liberal democracies.

The creation of good places in democratic societies thus requires balancing, in varying proportions, particularistic and universalistic ends. This goal reflects actual relations between community and place that may be described as an often uneasy mix of parochial attachments and cosmopolitan ideals. Both the normative and the actual point toward the usefulness of a more cosmopolitan conception of place than has traditionally been found in social and political theory. Such a concept remains rooted in the concreteness of everyday experience and practice but opens to the potentiality of a common humanity striving to make the earth a better home (Tuan 1996; Dewey [1916] 1944, 207–18). Conceived in this way, place enriches theoretical discussion of modern democratic community.

THREE

Transnational Networks and Political Participation

The Place of Immigrants in the European Union

RIVA KASTORYANO

An important number of networks—some formal, some informal; some based on identity, some on interest, some often on both—cross national borders and form a spiderweb that covers Europe. These networks are a part of a European space, defined by the Single European Act of 1986 as a "space without internal frontiers in which the free movement of goods, of property and capital is safeguarded." With the project to construct a political Europe, this space becomes a space where an array of transnational associations—cultural, political, ethnic, religious—combine their activities across national borders and express their quest for representation at the European level before the European Commission in Brussels or the European Parliament in Strasbourg.

Immigrants[1] with the status of permanent residents or legal citizens of one member state foster solidarity networks across national borders on the grounds of one or several identities, linking the home country to the country of residence and to a broader European space. Even though immigration and integration policies come within the power of the nation-state, such transnational associations seek recognition from supranational European institutions as loci of collective identity. The emergence of transnational associations underscores the development of multiple transspatial interactions: between national societies (home and host) and the wider European space; between national and supranational institutions; and among member states of the European Union. These multilevel interactions create common social, cultural, economic, and political involvement.

Transnationalism as an influential mode of action and Europe as a new political unit raise the same question: is the nation-state still relevant? If we define nation-states as political structures "invented" in eighteenth-century Europe based on the coincidence of territorial, cultural, linguistic, and even, to some extent, religious unity, then new global structures, such as supranational institutions and transnational networks, challenge them (Tilly 1974). Supranational institutions and transnational networks are at the core of the process of Europeanization *and*, on a larger scale, of globalization. Supranational institutions impose norms, values, and discourses on nation-states. Transnational organizations create a space for political participation that goes beyond national territories. Together they remap a political community that is Europe, albeit transnational and therefore deterritorialized or reterritorialized, or both. From this perspective, territory becomes a broader and unbounded space where nation-states and supranational institutions interact and where transnational networks build bridges between national societies and Europe. Many questions with regard to membership arise from these developments: what becomes of the relationship between citizenship, nationality, and identity; between territory and the nation-state; between rights and identities, culture and politics, states and nations — all concepts that are interconnected in nation-states?

This chapter examines the transnational organization of immigrants settled in different European countries. Based on the results of research that I conducted on transnational solidarities in Europe of immigrants settled in different European countries and involved in building transnational networks at the European level,[2] this chapter analyzes the multiple interactions between transnational networks, nation-states, and supranational institutions and their role in the emergence of a European space. It explores the role of transnational participation in defining bonds of transnational solidarity as well as European identity and citizenship and examines the relations between transnationalism and the nation-states. I argue that the transnational organizations, rather than contributing to the erosion of the nation-state, serve to redefine political structure and to redistribute the balance between nation and state, where the state becomes the driving force behind the construction of global structures and the nation serves as a source of mobilization. Supranational institutions, by encouraging such structures, promote a European space while reinforcing paradoxically the role of the state in the political construction of Europe, and of the nation as a source of identification. But at the same time their norm-imposing approach to politics leads nation-states to institutional changes and to a redefinition of the balance between state and community beyond territorial changes. Their approach affects the very concept of political community and European identity.

■ Transnational Solidarity and Identity: The Case of Association

In the early 1990s, more than 13 million "foreigners" (non-Europeans) were living legally in the twelve countries of the European Community. Sixty percent of the foreigners in France and 70 percent in Germany and in the Netherlands are citizens of countries outside the European Community. Of this group, France has absorbed most of the North Africans (820,000 Algerians, 516,000 Moroccans, and 200,000 Tunisians), and Germany has taken the largest number of Turks (almost 2 million). In the Netherlands, the Turks (160,000) and the Moroccans (123,000) constitute most of the non-European immigrants, while Great Britain is characterized by the preponderance of groups from India (689,000), the West Indies (547,000), and Pakistan (406,000) (OECD 1992; Statistical Office of the European Communities 1992).

These groups are increasingly organized into transnational networks. They take refuge in solidarity expressed by common nationality, ethnicity, and religion that cuts across boundaries in Europe, which they perceive as a new political space for collective claims and representation. At the signing of the Maastricht Treaty in 1992, when the European Union was constituted of twelve member states, some activists among immigrants involved in transnational networks expressed an explicit desire to go beyond national frameworks and spoke of "a 13th population" or "a 13th state" or even "a 13th nation." This formulation suggests a feeling of collective belonging through transnationality and a will to consolidate their solidarity as political communities within member states. Their organization and membership goes beyond territorial settings, creating a new transnational space (Faist 1998). The "13th" idea also points to the emergence of a transnational community on a European level, that is, a community structured by individuals or groups settled in different national societies, sharing common references—territorial, religious, linguistic—and defining common interests beyond boundaries. In the case of immigrants, voluntary associations constitute an avenue for such transborder organization and activity.

The emergence of transnational communities is a global phenomenon. Groups and institutions are involved in structuring networks based on economic interests, cultural exchanges, social relations, and political mobilizations. Increasing mobility and the development of communication have contributed to the intensifying of such transborder relations through constructed networks and organized communities. The institutionalization of a transnational community requires a coordination of activities based most of the time on common references, objective or subjective, and common interest among members. It also requires a coordination of resources, information, and sites of social power across national borders for political, cultural, economic, technological, and social purposes (Held et al. 1999).

Studies on the emergence of transnational communities emphasize the postcolonial immigration and the individual, commercial, institutional (political, cultural, and social) relations that immigrants entertain in the two countries. Operating on two countries gives rise to new practices and symbols.[3] In most of the cases, transnational communities are built on common geographical, cultural, and political references, hence their relative homogeneity as well as the intensity of intracommunal relations and the efficiency of their action.

The European Union presents a type of political space that differs in fundamental ways from national political spaces and induces a new type of transnational community. In the context of the European Union, transnational community transcends the boundaries of the member states, relating a vast European space, which includes the member states, to the country of origin of the immigrants. The emergence of European space is linked to multiple and complex interactions between states and collective identities expressed by immigrants or any kind of interest group that tries to imprint its independence on the state. Transnational actors such as leaders of voluntary associations, business persons, or activists develop strategies that extend beyond nation-states by expressing their solidarity through transnational networks based on a common identity or interest, often both.

Some transnational networks are based on local initiatives, some come from the country of origin, and some are encouraged by supranational institutions, particularly by the European Parliament. Initiatives on all levels help activists develop political strategies and mobilization beyond states. An association leader in Marseille stated during an interview: "We have to get used to addressing supranational organizations, to achieving an organization in Strasbourg or Brussels, that will be European and will cut through."

Such a community is far from being homogeneous. Immigrants from some geographic and national areas are more involved in building transnational networks than those from other areas. One of the reasons is that the colonial past has affected the immigrants' trajectories in different countries in Europe, leading to local concentrations that influence their motivation and involvement in building transnational networks. For example, Algerians are less involved in building European transnational networks. Because of the Algerian colonial past, their main reference remains France. Their understanding of transnationality is limited to the relation between France and Algeria and, to some extent, to all of North Africa. Their transborder networks are mainly elaborated around economic and cultural activities, with an obvious effect on social relations and political expectation on both sides of the Mediterranean.

A specialist in European immigration policy within the European Commission informed me during an interview that the Turks are "over-represented in the Immigrant Forum." The influence of the past explains the difference. Even though privileged specific economic ties led the Turkish immigration to Germany, the historical indifference of the Turks toward the countries of immigration explains their dispersion throughout Europe. They seem better armed to build a transnational network crossing the boundaries of many states. But lately in Europe, colonial ties are no longer a determinant factor. Increasing mobility, standardized social relations, convergent politics of immigration, and globalized markets generate a sort of geographical indifference: Algerians are not only in France; Indians and Pakistanis are not only in Great Britain. Networks driven by market and social opportunities in different countries have replaced the colonial ties that previously determined the trajectory of immigrant groups.

Another factor that motivates immigrants to build transnational networks is the fragments of national, linguistic, or religious identity that are frequently repressed within the home national political framework because of the immigrants' status as a nonrecognized minority. Immigrants reappropriate these kinds of identity through "identity politics" in the country of immigration. One step further is to seek the recognition of these identity fragments at the European level in order to gain legitimacy both within European institutions and in their own state. The best examples are the Kabyle from Algeria, the Kurds from Turkey, and the Sikhs from Southeast Asia. Their mobilizations find support in European institutions that encourage "national and regional groups"[4] as well as "stateless" populations. It is within this framework that Kurds find institutional legitimacy in Europe. This procedure is the result of a codevelopment policy of the European countries with the countries of immigration. It comes down to including the countries of origin in the representation of the migrants in Europe and consequently provides a legitimacy of action and a claim to populations in a minority situation in their home country. Like the claims relating to rights of residence, citizenship, and protection from expulsion that are directed at European institutions, the interests expressed in terms of the identities of populations formed through immigration find opportunity for action within a political Europe under construction, thus leading to new forms and structures of representation and to new negotiations with their state of reference.

Thus, European supranational institutions, through transnational actions, play an important role in the diffusion of social, cultural, political, and even juridical norms in different European countries as well as in the country of immigrants' origin. Guided by the logic of regulation and of political and juridical harmonization that they impose on nation-states, European supranational institutions have encouraged a global structure and have moved forward to define a common platform for the network. They have also intervened in the definition of criteria on which such a community should rely and have helped the actors to find a common denominator to deal with claims on a European level, that is, beyond the relations with the nation-states.

Initially, social, economic, political, and legal motives corresponding to the main concerns of European public opinion — both immigrant and nonimmigrant unemployment (67%) and racism — legitimated the shaping of networks.[5] In 1986 these concerns led to the creation of the Forum of Migrants. The European Parliament mobilized resources for immigrants' voluntary associations in order to help them to coordinate their activities. Though dissolved in 2001, the Forum had initiated a transnational structure, a sort of European federation of immigrants' associations. According to the person in charge of issues related to immigration in the Commission, this structure owed its existence to the financial policy of the European Parliament. The declared objective of the Forum was to create a place where non-European immigrants could make their claims known and also share their demands. They could, furthermore, circulate the information that European authorities produced about them (Neveu 1994, 95–105). According to the president of the Forum (which was linked to the Commission in Brussels), the intention was that nationals from non-European third countries who had settled in member states should enjoy the same rights and opportunities as indigenous citizens of those states. The Forum also wished to make up for any democratic deficits. The explicit goal was to fight against racism with a common jurisdiction in different European countries.

The criteria for voluntary associations to be a part of the Forum were defined by the European Parliament. These associations had to be supported by the welfare state of each country and therefore recognized as legitimate. They also had to prove their capacity to organize (locally and/or nationally) and to mobilize human and material resources; to define their activities as universal (based on universal values such as equality and respect for human rights); and to speak on behalf of populations coming from non-European countries. The Forum was thus formed of voluntary associations that were already recognized in their country of settlement. Measures of the organization's effectiveness included the multiplicity of nationalities in it, the number of branches, the extent of its networks, the plurality of sectors they covered (economic, social, cultural), and how representative they were in the countries where they were located.

A network built on a common interest, defined at the European level and formulated in terms of equality of rights, was meant to liberate immigrants from the politics of their home country as well their host country and to express claims beyond both nation-states. As for their leaders, they developed a discourse on equality and the universality of human rights, seeing the transnational effort as a way to fight racism and xenophobia globally. But they had trouble finding a common base to coordinate their activity at the European level. National particularities emerged in official rhetoric and political claims. Their rhetoric echoed collective identities that had been shaped in relation to their states of residence (Kastoryano 2002). Interests also developed and were formulated in reaction to states' policies on immigration and on integration. This affected claims expressed by different identities that they would like to be recognized by supranational institutions, in reaction to national identities.

The interaction between states and immigrants led the immigrants to define a core identity from which a community could be constructed in order to negotiate its recognition with the state (Kastoryano 2002). National specificities and political traditions that immigrants had internalized through their participation in voluntary associations in the country of settlement were carried to the European level to negotiate with supranational institutions for recognition. In France, for example, the republican rhetoric on citizenship and a defensive discourse on secularism called *laïcité* have led immigrants to claim the recognition of a religious community within state legitimacy. They base their claims on religion. In Germany, Turks claim the recognition of an ethnic minority based on a common foreign nationality, expressed by their claim for dual citizenship, where citizenship is expressed in terms of rights and nationality as an ethnic identity (Kastoryano 2002). The emphasis is then on common nationality. In Great Britain, Caribbeans, Indians, and Pakistanis aim to overcome discrimination in terms of color.

On the European level, discourses demarcate national particularities in reaction to other national and political contexts. British activists, for example, reject the term *immigrant,* considering it inappropriate to their situation. They want to see a law on the "equality of the races" appear in Europe as in Great Britain; a U.K. organization named SCORE (Standing Conference on Racial Equality) emerged in 1990 to express publicly the fear of a Europe that would define its own identity by exclusions, not only in terms of the legal status of foreigners, but also in terms of race and racism. Leaders of voluntary associations in France reject, in their discourse, any policy dealing with *ethnicity,* a concept that they feel is relevant for the British context but not for the French one. They are expressing their attachment to French rhetoric, according to which policies toward immigrants are to prevent a social exclusion from the larger society and not to recognize a cultural specificity. Black identity developed in Great Britain and ethnic identity expressed in terms of nationality in Germany have become a way for activists to fight against racism and discrimination and for equality of rights in Europe. Therefore, collective actions in these countries stem from the fight against any kind of exclusion, social, cultural, or political.

But at the same time, "immigrants," "foreigners," or "Blacks," according to the terminology in each country, all converge in their political strategies and participation in different countries. States define themselves as republican and assimilationist, like France, or exclusivist in terms of citizenship, like Germany, or they promote the formation of racial communities in the public sphere, like Great Britain. In many European countries, immigrants develop strategies based on a collective representation of cultural, national, or religious identities. Thus the main criteria appear to be identity. Identity of "origin" or "identity of circumstance," as Leca (1992) puts it, but in any case identities constructed and defined in relation or in reaction to each nation-state constitute the links of the chain.

The Forum of Migrants established a somewhat ambiguous criterion of membership for associations. It was dependent on the "nationality" of an association's leaders as well as their candidates. This ambiguity was rooted, in turn, in defining an "immigrant" and in particular an immigrant's legal status in the country of residence. Who are these immigrants? Algerians in France? Most of them have French citizenship; the nationality of their leader refers in reality to a "nationality of origin" and not to a legal affiliation. In Great Britain, Indians, Pakistanis, and persons of Caribbean origin hold British citizenship. This criterion suits Turks in Germany, who are still "foreigners," and Africans. What about immigrants from Italy, Portugal, Spain, and Greece, who came from an EU member state other than the state of residence? They are citizens of the Union but must cope with racism and all the social baggage of immigration.

The issue, then, is not nationality (non-EU) but ethnicity. Behind the criterion of nationality that was considered to be a juridical one, and therefore objective, voluntary associations that were part of the Forum expressed a "nationality of origin," or a religion (mainly Islam) related to it, or color. But the declared goal of supranational institutions forces them to dissimulate identities in their claim for recognition as non-Europeans. So it became a matter of ethnicity, defined as a subjective feeling of belonging and to some extent of membership (Weber [1922] 1978, 395).

And Islam? If identity provides the cement of the networks, Islam is the hard core. Islamic associations use the European space in the same way as the cultural and social associations but—in the name of secularism—do not get any support from either national or supranational institutions. Representatives of Islamic associations work mainly in connection with the home countries or with the help of international organizations, or both. The home countries try to rally their nationals to achieve recognition of their (extracommunity) country from the European authorities. Thus they reactivate their loyalties through religion and contribute to the creation of a transnational community. The international organizations interested in Islam in Europe mobilize resources to allow Islam to go beyond the national diversity of Muslims living in the various countries of the

Union, to create a single religious identification and a transnational solidarity based on that. Because of this policy, those religious networks have fit into the European system and rival the sociocultural associations on a local level.

Yet, coordinating the Islamic networks in Europe may be even harder than coordinating the associations defined as social, because, although the Islamic associations are autonomous with respect to the welfare state of the various European countries, they relate like the so-called cultural associations to the public authorities of the countries of residence. "I try not to take a position that goes beyond the borders. We are in France, our goal is to defend Islam in France," said the leader of the National Federation of Moslems in France, in an interview. Similarly, in Germany—as in the Netherlands—the Islamic associations are part of the federations of associations grouped by nationality that are seeking representation as a minority in the country of settlement.

Another difficulty comes from the diversity of the nationalities, sects, and ethno-cultural groups among the Muslims in Europe. Some associational groups are concerned about European representation and present themselves as multinationals, collecting several nationalities of origin while branching out in the various countries of the Union, like the Jamaat-Tabligh (Faith and Practice) organization. This organization is of Indian origin and was first established in Great Britain in the 1960s. Since then it has extended its networks into France and Belgium, and recently into Germany and the Netherlands, sending missionaries into local communities to promote the faith of the Muslims and, obviously, to get them to support the organization (Kepel 1987). This movement "transcends not only material boundaries, but also sects, legal schools, and Sufi orders in their ideological conception," and its activists express the desire "to be good citizens," avoiding all political positions since, according to them, "politics divides Islam" (Diop 1994, 147).

Other Islamic associations, however, openly express the political position of Islam in the international system. Nevertheless, most of them remain confined to the nationality they represent, and especially to the political parties for which they serve as spokespersons in Europe. This is the case, for example, of National Vision, the affiliate of the successive religious parties in Turkey, and their organization, also called National Vision, with its twenty-eight foreign offices in Europe, ten of them in Germany. Opposing the consular network of religious affairs of the Turkish state (Diyanet), the organization's activists are trying to create a transnational solidarity based on a political identity expressed through religion which remains actually Turkish. Algerian networks of the Islamic Front of Salvation or those of its armed branch, the GIA, pursue the same objectives of political legitimation with their followers. If these networks are necessarily transformed into structures of [illegible] on behalf of the religious or political identification they share with Muslims of other countries, their presence affects both Europe and the home countries, especially overall relations between Europe and a Muslim bloc. The dissolution of the Forum does not seem to have affected the organization of Islam in Europe; on the contrary it might have tended more to legitimate the status of a religious minority and its transnational solidarity.

In short, Islam in Europe is seeking unity in diversity through transnational solidarity. But since the organizations claiming to represent it are generally outside the formal networks that existed in the Forum of Migrants, they have intensified their development of a system of solidarity based on religion. Yet, as with social and cultural networks, their strategy aims at the recognition of identities that are primarily national and ethnic. Despite the influence of the home countries or the international organizations that endow them with political importance, their claims are adapted to the European context. They also raise a question of representation in European institutions, especially since the European Convention on Human Rights (ECHR) recognizes freedom of religion. Article 9 of the ECHR states that

1. Everyone has the right to freedom of thought, conscience and religion; this right includes freedom to change his religion or belief, and freedom, either alone or in community with others and in public or private, to manifest his religion or belief, in worship, teaching, practice and observance.
2. Freedom to manifest one's religion or beliefs shall be subject only to such limitations as are prescribed by law and are necessary in a democratic society in the interests of public safety, for the protection of public order, health or morals, or the protection of the rights and freedoms of others.

■ European Identity and Citizenship

What are the implications of transnationality for European identity and citizenship? By selecting only associations representative of immigrants from third-country nationals, the Forum of Migrants excluded non-European immigrants from the incipient European identity and created a gap between European immigrants and non-European immigrants. Moreover, it created an ambiguity in the definition of European identity and European citizenship. A future can at least be imagined when the permanent presence of those non-European immigrants can create an official status of European resident with specific political rights. But for the time being, official European initiatives have been producing only a minority in Europe. This minority is defined by a current or original non-European nationality, and its representation by the Forum of Migrants assumed an ethno-religious aspect. It therefore reflected an exclusive idea of European citizenship, just as nation-states define the national citizenship. To some extent, it reinforced the nationalist sentiments expressed by the member states confronting the European construction.

Debates on European citizenship bring to the fore the disruption of the nation-state model as well as the difficulty of parting with it. Many concepts, such as postnational, cosmopolitan, or transnational membership, were elaborated along with the transformation of the European economic community into a European union, when in 1992 the Treaty of Maastricht marked the first steps toward its unity. The French philosopher Jean-Marc Ferry (1991) underlines the taking over of the "nationalist principle" fostered by the building of a political Europe; he suggests a postnational model, a membership beyond the nation-state. The German philosopher Jürgen Habermas (1992) develops the concept of "constitutional patriotism" in order to emphasize the need for a "cosmopolitan solidarity" that implies the separation between the feeling of belonging that national citizenship involves and its legal practice beyond the framework of the nation-state. Taking into account the non-European populations deriving from immigration in the 1960s, Soysal (1994) defines as postnational the fostering of international norms referring to the person or residence instead of legal citizenship.

All these approaches express a normative view of citizenship that European political projects do not necessarily follow. Europe is being built with supranational institutions that oppose in form and content the postnational concept. While postnationalism would lead to a recognition of cultural diversity and the acceptance of pluralism as the basis for European belonging, the supranational approach taken in the construction of a unified European space mimics the nation-state. In 1992 the Maastricht Treaty defined the status of citizenship as "citizenship of the Union" (Art. 8). A "citizen of the union" is whoever holds the nationality of one of the member states. The "citizenship of the Union" requires national citizenship in one of the member states. Thus the treaty maintains the link between citizenship and nationality, as is the case in nation-states. But the practice of citizenship in the Union (direct participation: vote) brings an extraterritoriality with regard to nation-states. Again, Article 8 (8a–8d) of the Treaty of Maastricht gives the citizen of the Union the right of free circulation, the liberty to reside and work in the territory of a member state, and even the right to vote and to become a candidate in local elections and in the elections of the European Parliament based on residency, that is, on the territory of a member state of which he or she is not a citizen, but just a resident. As Preuss (1998) has pointed out, territoriality becomes the basic means of citizenship in the Union. But it introduces at the same time an extraterritoriality to the concept of citizenship, extending its practice beyond territorially limited nation-states, therefore deterritorializing the national community or reterritorializing the European space.

That is precisely the locus of transnational participation and of the formation of a European citizenship. On a national level, the increasing political participation of immigrants through voluntary associations has contributed to the formation of an identity of citizen in the country of residence. This identity has been shaped in relation to national institutions that have created an identification with the political community through collective action. On the European level, the identity of citizen is being shaped in relation to supranational institutions, which make Europe a public good that generates new political identification for individuals involved in transnational mobilization. Their political acculturation on a national level has become a "passage obligé" for a political engagement on the European level. Such participation can be considered a second stage of political socialization for immigrants and the European space, a political space where they exercise citizenship beyond the political territories of the state. From this perspective, leaders of immigrants' associations, legal citizens of a member state or not, act together in new space, making it a common space of political interaction and power.

A new discourse about citizenship accompanies the immigrant activists' political commitment to equal rights and to the fight against racism on the national and European levels. As one association leader argued, "We are European citizens; we are part of the European landscape." This activist sees the European landscape as a space formed by a spiderweb of networks of solidarity and interests spread over a territory that now covers fifteen countries. Immigrants active in the Forum maintained that their presence in this web through transnational networks conferred a "right" to participate in shaping Europe and introduce activists to "European citizenship." From their perspective, citizenship derives principally from political participation in public life. The engagement of individuals in politics and their direct or indirect participation in the public good is an expression of citizenship (Leca 1986).

Such political identification leads to confusion in the definition of legal status with regard to the couple citizenship/nationality. For immigrants with a non-European background, European citizenship underlines the complexity of their reality. The concept *European citizen* is fraught with paradoxes. By stimulating immigrant involvement in the "common good" that the European Union represents, supranational institutions extract immigrants from their "primordial ties" by taking them away from any direct political action within their home and host countries and bringing them into a common identification defined by a common interest that is European. But paradoxically, European citizenship, as a more global concept of membership than nation-state citizenship, introduces the allegiance of immigrants to their home country into the process of bargaining in the same way as it expresses their allegiance to their state of residence and to the transnational community in which they are involved. Citizenship is then conceived as Habermas (1996, 431–514) suggests, as giving a legal position to an individual outside the nation-state.

Multiple belonging that is related to transnationalism appears in the way the individual, the citizen, combines and classifies his loyalties. As the leader of the Association of Moroccans in France put it: "Today we must no longer consider ourselves 'Moroccans,' but as belonging to the European community, Frenchmen of Moroccan origin." He could have added "of the Muslim religion" without threatening either French *laïcité* or secular Europe, but simply expressing an attachment to a religious culture that, like "Moroccan origin," conveys attachment to a tradition, a history, and of course, a religion. Those of Algerian, Tunisian, Turkish, or any other origin that brings out the cultural diversity within nation-states, might also be part of a "minority belonging to the European community": an ethnic national and not a regional minority determined by the norms of the European Council, or a Muslim religious minority in a Judeo-Christian and multiconfessional Europe.

Europe conceived as an open space offers a multiplicity of identifications and loyalties to transnational actors. Transnationalism as a new mode of participation helps to assert the autonomy of these identities and of their representatives in territorially defined nation-states while it fosters immigrant involvement in the "European project." Just as in the United States, a country of immigration ever since it was founded, where the various waves have contributed to defining the American nation, in Europe, the European Parliament feeds hope to non-European immigrants that they will participate in the construction of Europe and its identity. Transnational associations rely upon Europe as a new political space open to all kinds of claims and representations because of its uncertain or "soft" identity in contrast to the "hard" national identities. Nation-states and national identities, products of a long process made of common experiences rooted in history and collective memory, are experienced by immigrants as difficult to penetrate. In contrast, Europe, perceived as an uncertain political community, might leave a space for collective actions and claims through which plural and complex loyalties can be articulated in order to define a new political unity and identity. The hope, in sum, is that a "plural" or "multicultural" Europe might be fashioned (Kastoryano 1998).

European nationals as well as non-nationals and "minorities" and leaders of voluntary associations already affirm that Europe represents "the will to live together." They express a responsibility to shape the new "community of fate." By using these formulas, the first by Ernst Renan and the second by Otto Bauer,[6] they reproduce the rhetoric of the political class and of European supranational institutions, a rhetoric that evokes the ideals behind the creation of nation-states, as if nation building could be a model for the construction of a political Europe. While the process of nation building occurs with the creation of institutions and the production of knowledge, the construction of a political Europe follows the opposite path: European institutions "show the way" by producing a discourse, defining the modes of organization, and operating on nation-states with the same objective of unification that national institutions have for national societies (Fabre 1996). Guided by the project of regulation of national political traditions and juridical harmonization, the European institutions impose such norms on nation-states in the name of the "general interest" that is "Europe building."

But to what extent does the institutional construction of Europe produce a unified European political space that reflects the general will, one that translates into solidarity among citizens of different member states and between citizens and residents? The Forum of Migrants showed that immigrants have reproduced the discourse they internalized in their fight for equality and against racism and have claimed the recognition of an identity they constructed in relation to their state of residence or formal citizenship. With regard to the Indonesian nation, Geertz notes that "some of the most critical decisions concerning the direction of the public life are not made in parliaments and presidiums; they are made in the unformalized realms of what Durkheim called 'the collective conscience'" (1973, 316). In the European Union, the discourse on the general will by national political classes in competition for power in Europe did not seem to find an echo in the definition of a European collective conscience among immigrant nationals as non-nationals. Charles Taylor is correct when he says that "[i]t is hard to conceive of a democratic state that really lacks every identifying dimension" (1992, 148). Obviously the Forum could not set up a solidarity initiated by institutions that mobilize resources without being able to create an identification with Europe. The Forum's dissolution left space for the informal network built by immigrants themselves, which is thus not recognized by the European Parliament but is successful in developing a collective conscience of being a minority.[7] The identity of minority found its ground in religion such as "Muslims in Europe." Although most of the immigrants from a Muslim background define themselves as secular, the absence of religion from the political projects of the European Union has led religious organizations to consolidate their transnational networks. The various members, even if they express their loyalty to different nations and even to different states, have developed a common identity based on the experience of being Muslim in Europe and have redefined solidarities within European space so that they claim a representation and recognition at the European level.

Another question, related to the previous one, has to do with the nature of the European space and its capacity to reconstitute a political community. For Dominique Wolton (1998), multiple networks lead to a symbolic space that provides interconnection and communication among various social activities but does not generate a public debate. As a matter of fact, on the one hand, the claim of representation has limited the relationship of immigrants to European supranational institutions and has engaged a small fraction of the immigrant population: the leaders of some selected voluntary associations. Their involvement and claim has produced no echo whatsoever in public opinion—national or European. On the other hand, the leaders carried to the European level their specific relations with their own states of residence (and of legal citizenship in some cases) and claimed the representation of their own interest defined within the process of negotiations with them (Kastoryano 2002).

Nevertheless, the very nature of Europe, where various networks (communal, national, regional, and religious, as well as professional) compete and interact, has generated a European civil society that is transnational. The European civil society remakes territory as a form of national corporatism. A corporatist civil society where interest groups are organized in associations constitutes a tool for activists to formulate new demands and puts the voluntary associations in relation to a wider space. Supranational European institutions play an important part in the formation of a European transnational civil society. They encourage interaction among voluntary associations and help to make visible various fragments of identity represented by these associations on the European level. According to Article 6 of the Amsterdam Treaty, "The European Union should respect the national identities of its member states." According to this logic, fragmentation within the European space is limited to national identities and does not reflect internal diversification of national societies. This logic also considers the internal diversity of the nations (regional, linguistic, ethnic) "acculturated" or even "assimilated" (Bauböck 2000). Such a transnational space is insufficient to create what Ferry (2000) calls a "social link." In fact, by creating the Forum, European supranational institutions played the same role as welfare state institutions on the national level, promoting identity politics with the same paradoxes and the same unexpected consequences but without the same function of national integration and without being able to create European-wide civic or ethnic solidarity. The participation of the leaders of voluntary associations in European transnational civil society involves them in the multiple interactions and confrontations of the cultures that form the European Union. Their participation affirms on the one hand a space for rational action, political and social development, and on the other a space for cultural integration of a collective identity. Politicization on the European level situates the immigrants' claim toward both states and the European Union.

■ Transnationalism and the Nation-State

The paradox of supranationalism as an approach to the construction of a unified Europe lies in the fact that while questioning the nation-state, it also reinforces the role of the state in the building of a political Europe. On the one hand, supranational institutions challenge nation-states. By creating the Forum of Migrants, the European Parliament marked its autonomy from national institutions and induced immigrants to situate themselves beyond nation-states. On the other hand, using the same criteria as the welfare states of national institutions, the European Parliament projected collective identities that are questionable on a national level onto a European level. European space has become the projection of the nation-state onto a transnational scale. On the national level, people facing social problems because of their origin use their homeland identities and nationalities to initiate fights against racism. Transnationality is also fraught with paradox. The consolidation of a transnational solidarity generally aims to influence the state from outside. Even if transnational networks contribute to the formation of "external communities" out of their relationship with states, these networks today are imposed on the states as indispensable structures for negotiation of collective identities and interest with the national public authorities who define the limits of their legitimacy. The objective of a transnational network is to reinforce its representation at the European level, but its practical goal is recognition at the national level. In addition, activists, even the most active ones at the European level, see states as their only adversary. The difficulties that voluntary associations have in coordinating their actions and their claims when they spring from their own initiative, without the intervention of supranational institutions, underscore the heavy hand of the state.

In other words, the ultimate goal is to reach a political representation that can only be defined at the national level. Rights and interests for non-Europeans—such as the protection of their rights as residents, housing and employment policy, family reunification, and mobilization against expulsion, in short policies that touch, directly or indirectly, the domain of identity—can only be claimed from the state. But from now on, all claims at a national level imply a parallel pressure at the European level, and conversely, all claims on the European level aim to have an impact on decisions taken on the national level within each of the member states. As the leader of the Associations of African Workers in France put it, in an interview: "For us, immigrants from the third world, we must act in such a way as to be in an effective position to get organized and protect ourselves, to carry our claims high; since the bulk of our recommendations which are backed up by the EEC and often favorable to us are not always seen in the best of light by the member countries. . . . Let us act in such a way that what is positive at the European level be echoed in the country of residence."

A united Europe introduces a "normative supranationalism" (de Witte 1993) that it imposes on nation-states. Even if the issue of human rights, for example, remains the exclusive jurisdiction of the states, they are forced to accept the new legal norms produced by European institutions, since the European Convention on Human Rights entitles the citizen of the Union (in this case, one who has the nationality of one of the states that has accepted individual appeal) to apply directly to the European Council and entitles a foreigner (who does not have the nationality of a member state of the Union) to appeal to the European Court of the Rights of Man. In case of deportation, for example, the foreigner can oppose the national decisions on behalf of the right of respect for family life (para. 1 of Art. 8), after exhausting the paths of internal appeal. Thus, the principle of the Convention is based on the idea that "the individual, previously isolated and ignored in relations between states, becomes a person, a citizen in the community of European nations."[8]

Attorneys are also considering a classification of rights: civil and political rights, first; followed by economic, social, and cultural rights; and finally, "rights of solidarity" (Sudre 1995, 153–58). Rights of solidarity refer to freedom of collective action in a community framework and declare that it is "only in the community the full development of the individual personality is possible" (154). Introducing this right of solidarity thus emphasizes the ambivalence between individual right and collective right, a new concern of states and groups that appears not only in public opinion but also in political discourse.

Since the right of solidarity for immigrants might refer to minority rights, it presents a challenge to nation-states. In the European context, the concept of minority has been developed since the fall of the Berlin Wall in reference to the social, cultural, and political reality of the countries of central and eastern Europe. Since 1989, the problem of democracy in those countries has been posed in terms of recognizing minorities; it has generated the application of minority rights in European institutions of countries of western Europe. Yet this concept is so ideologically charged that it is hardly portable. For example, on 10 November 1994 the European Council developed a framework agreement guaranteeing the individual freedoms of minorities without infringing the unity and cohesion of the state. France did not sign the agreement because the French minister for European affairs judged the text "incompatible with the Constitution."[9] According to the same principle, France rejected also the recognition of regional (minority) languages recommended by the Charte des langues régionales.

Yet, the term *minority* is ambiguous. Does it designate cultural, linguistic, territorial, and officially recognized minorities, such as the Catalans and the Basques in Spain and the Bretons and the Corse in France; or does it refer to minorities composed of immigrants and also officially recognized as immigrants, as in the Netherlands? According to the European Convention on Human Rights, "the word minority refers to a group inferior in number to the rest of the population and whose members share in their will to hold on to their culture, traditions, religion or language" (Art. 29, para. 1, of the ECHR). In France, whether it concerns regional or religious identities or collective identities that are evident within populations born of immigration, the term is being rejected. In Germany, it refers to German minorities only, settled outside German territory. Turkish nationals in particular draw inspiration from the official usage of the term when they demonstrate the will to structure a Turkish or Kurdish national community in Germany. The rearranging of their associations in this sense drives the Federal Republic to react in similar terms. Each country develops arguments dictated by what it considers to go against its national integrity.

Appadurai's "theory of cascade" (1996), which links the local to the global, appears in the European context as the result of increasing interaction between nation-states and supranational institutions in the definition of general norms and values, while a national particularity is retained for each state—particularly when dealing with policies on immigration, integration, and access to citizenship. Supranationality raises tension between European institutions—intergovernmental relations and nation-states—when it is a matter of asylum, immigration, or integration policy; the tension is between a tendency to unify a European political arena and states' sovereignty.

The European Union stands for the idea of open-minded conciliation—for a conception of universality that contrasts with that of the nation-state. According to those who defend immigrants' rights, this universalism conceives Europe as an arena in which foreign residents and even citizens who are perceived as foreigners (due to national origin, color, or religion) can live together. To imagine a transnational community born of immigration would give support to nationalist sentiments voiced by the member states facing immigration on the one hand and the building of Europe on the other hand. At the same time, the irrationality of national sentiments amounting to no more than ethnic belonging stands opposed to the rationality of the European institutions that define legal norms in such areas as human rights and the rights of minorities in particular, areas that concern the "internal foreigners." Of course, an organization that transcends national borders brings to the fore the principle of multiple identifications deriving from the logic of a political Europe. It is precisely this aspect of multiple identification and allegiances that provokes passionate debates about the construction of Europe, for it disrupts the relations between citizenship and nationality, states and nations, culture and politics, as well as the relations between a political community and the territorial nature of participation. It signals therefore the nonrelevance of the nation-states and its unitarian ideology facing identity claims being expressed within and without national borders.

But a transnational organization of immigrants in Europe is a sign of the Europeanization of a political action, but not of the Europeanization of claims (Tarrow 2001). Claims for recognition and equality remain focused on the nation-state. On the one hand, the nation-state still provides the basic framework within which negotiations occur, and it fashions the legal and institutional setting for recognition. On the other hand, the nation remains a source of identity and a catalyst for emotions.

Therefore, the empirical evidence suggests that states remain the driving force of the European Union. Even if they submit to supranational norms, states keep their autonomy in internal decisions, and in international relations they are the main actors in negotiations. They concurrently accede to supranational norms and maintain their autonomy. The nation serves as the basis of any transnational enterprise. Therefore, the permanence of the nation-state as a model for a political unit in the construction of Europe relies very much on its capacity to negotiate within and without, that is, its capacity to adopt structural and institutional changes to the new reality. This development pushes states to imagine themselves as nonstate transnational actors, coordinating their interests and their strategies beyond their territory. Within this context, Europeanization means globalization on a smaller scale.

NOTES

1. The use of the word *immigrant* needs an explanatory note. Within this context *immigrants* means third-country nationals who settled in different European countries in the 1960s mainly for economic reasons. In many cases, they came from former colonies. Legally, *immigrant* refers to a temporary status, which is not valid today since most of them have the citizenship of the country of settlement. The use of the term, with these different meanings, reflects the difficulty of admitting that these populations are a part of the social, cultural, and political system of the country of residence.

2. This research was financed by the French Ministry of Research and conducted within the Centre d'Etudes et de Recherches Internationales (CERI) in Paris (1992–94). The results have been published in a special issue of *Revue Européenne des Migrations Internationales,* titled "Mobilisation ethniques: Du national au transnational" (Kastoryano 1994). I thank Catherine Neveu and Moustafa Diop for having participated actively in the field work as well as in the final analysis.

3. See, e.g., Basch, Schiller, and Blanc 1997; Cohen 1997; Gupta and Ferguson 1997; Hannertz 1996; Portes 1996; Levitt 1998.

4. The phrase refers to nationals of member states and of third countries, respectively. For example, the regions are defined as the Maghreb (North Africa), sub-Saharan Africa, Latin America, the Caribbean, and Turkey. Although the regions are defined as groups of countries, reflecting their geographic and cultural proximity, Turkey is represented as a region on its own. The Kurds, who can only be identified through their self-definition, represent around 30 percent of the Turkish immigrants, divided proportionally over the various European countries.

5. According to the publication *Eurobaromètre,* December 1992, in 1990, 29 percent of the individuals surveyed wanted to curb the rights of immigrants. In 1992 that rate had risen to 34 percent. Similarly, in 1991, 60 percent would have accepted immigrants from the Mediterranean with restrictions, and in 1993 only 46 percent of the population would have done so.

6. For Otto Bauer, the nation is the totality of men bound together through a common destiny into a community of character through a "common destiny" (1924).

7. The official reason for its dissolution is related to the use of the financial resources allocated to the Forum by its representatives, which does confirm the difficulty of establishing European solidarity based on equality.

8. Statement of François Mitterand quoted by Louis-Edmond Petit in the preface to Berger 1994.

9. *Le Monde,* 21 March 1995.

PART II

Political Organization, Culture, and the Problem of Scale

FOUR

National Identity Repertoires, Territory, and Globalization

ROY J. EIDELSON
AND IAN S. LUSTICK

Events and initiatives on the Continent are propelling individual states, or at least many inhabitants of these states, toward the adoption of a shared European identity. As a result, the dynamic aspects of previously secure links joining territory, political affinity, and culture are brought into view. While the thorny issues posed by globalization and its by-products are indeed being debated and confronted around the world, efforts to remap Europe into a supranational entity with common currency and open borders are especially intriguing because they represent the clearest attempts to design, encourage, and control these transformations. The authors of this chapter welcome the opportunity to contribute to the scholarly exchange on this topic by offering a very different research perspective—an agent-based modeling approach—inspired by complexity and evolutionary theory and employing computer simulations and Lustick and Dergachev's Agent-Based Identity Repertoire (ABIR) model.

At the heart of the ABIR model is the central constructivist notion that identities are best conceptualized as repertoires of possibilities. That is, identities emerge from multitudinous interactions among carriers of different ways of being, different performance styles, or different psychocultural stances (depending upon your disciplinary or epistemological commitments). At any level of analysis, the identity carrier (individual, group, political community, etc.) holds within it a subset of all available identities—its repertoire. Constructivist theory envisions these repertoires as offering a range of fluidity and malleability for identity carriers facing a world likely to encourage activation or self-presentation on a particular identity at one time and another identity at a different time. This activated identity substitution occurs without great difficulty when the substituting identity is part of the individual's repertoire. Change in the composition of the repertoire itself—that is, in the set of readily available ways of being (for an individual or for a group)—is also possible, albeit only in response to relatively large outside pressures and incentives.

By setting out this now-dominant approach to identity construction and reconstruction in rather more abstract and more precise terms than is customary, we can proceed to elaborate and apply the intuitions of constructivist theory. In this chapter we seek to apply this approach to questions of what it currently means, in a national political context, to be a member of a certain national community by illuminating the fundamental relationship between location (territory) and population (the mix of natives and immigrants among the inhabitants of territorially defined states). We are specifically interested in studying these questions as they are presented under conditions of globalization that render borders between states more porous and that tend to increase proportions of immigrants within formerly more homogeneous national territories.[1]

By applying these concepts to Europe at the dawn of the twenty-first century, we can move beyond objective and dichotomous designations encouraged by measuring individuals simply on the basis of habitation, citizenship, or current identity performance. Within our model, the full spectrum of identities available in the world includes a subset of national identities. Thus, in any particular national territory or country, there are a variety of different ways to be German, Spanish, or British, though there are, in the world as a whole, many more ways of being non-German, non-Spanish, or non-British. For example, an information-age young Green-oriented Berliner expresses Germanness in a way that is different from (but equally as real as) that of a Bavarian *brewmeister,* a neo-Nazi skinhead in Hamburg, or a Daimler-Chrysler executive in Bonn or Munich. And each of these individuals may have, in fact, the personal resources to perform his or her Germanness in one, some, or all of the other ways Germans do so. The same could be said about British identity with reference to a London barrister, a Cockney football fan, a Pakistani shopkeeper in the Midlands, a Welsh pony-trekking guide, or a Cambridge don. The extent to which these different identity stances are to be considered German or British may change over time, but it is important to emphasize that there are certain identity styles, or ways of being, that at any particular time or period do not fall within these categories. Thus, a Venetian gondolier or a Paris taxi driver would likely not be able to present himself credibly as German or British. Similarly, should a Briton or a German try acting as a gondolier or a Paris taxi driver, that identity would be understood as a put-on, as an act, and not as a real performance.

In short, individuals affiliated with a national state have identity repertoires that contain certain elements and exclude others. The different native ways of being French are variations on the theme of being French. There may also be identities in France that are not native French identities. In our simplified model, these non-native identities, which are nevertheless held in the repertoires of many inhabitants of France along with native identities, may be thought of as naturalized or immigrant identities. They may be seen as French, but neither as native (identities present only in France) nor alien (identities present only outside of France). At the extreme we may say that if an inhabitant of France holds no identities in her repertoire that are present only in France, then she could be considered an alien, not a naturalized French person. By extension, if no inhabitants—or only a minority of inhabitants—of France any longer hold identities in their repertoires distinctive to this territory, then we may say, indeed, that what was France is no longer. It is, then, differing identity repertoires, more than possession or performance of any one identity, that ultimately establish one person as French or Spanish and another person as Dutch or German and make one country France and another not France.

If identities did not arise via constructivist processes from distinctive but impermanent repertoires (e.g., if identity was instead automatically produced by some single unchanging factor), then patterns of identity politics would be much simpler. For example, if geographical location was the single determining factor, then the moment an individual emigrated from one country to another, he or she would immediately be indistinguishable from others in that country. This is clearly not the case, as the long-standing and now increasingly heated furor over issues of immigration demonstrates. At the other extreme, if identity is deemed to be purely a function of biology or early childhood upbringing, a Turk will be a Turk regardless of how long he or she works in Germany or how much he or she tries to be German. In sum, it is only because identities inhabit changeable repertoires that ways of being can move across a territorial or bodily border so that authentic Irishmen can adopt lifestyles or stances associated with Silicon Valley or the denizens of Parisian cafés.

Thus, constructivists are led to ask questions such as, How many non-German identities can be present within an individual before that individual is to be considered non-German? or How many individuals with incompletely German identity repertoires can there be before Germany is changed into something it most definitely was not? Such inquiries are posed with powerful and even explosive effect by waves of non-European immigrants entering Germany, Italy, Spain, France, Britain, and other European Community countries from Turkey, the Balkans, North Africa, sub-Saharan Africa, eastern Europe, and South Asia. These are challenging questions for study using constructivist theory, but they cannot even be posed using primordialist categories.

Our research explores these issues by modeling what happens as societies allow into their borders significant numbers of individuals with identities in their repertoires that previously did not exist within that society. What happens, in other words, as increasing proportions of the inhabitants of Germany share only a decreasing proportion of native German identities in their repertoires? How much alienness can be accepted or welcomed—either because resistance is insufficient or because alternative identities are now seen as having certain key advantages that merit their inclusion—before the status of German as the dominant identity within the territory is endangered? And what happens to the competition among German identities when that whole group of identities is challenged by naturalized or immigrant identities? Are there trigger points beyond which the much-discussed phenomenon of *glocalization* or "reactive nationalism" suddenly arises as a potent expression of parochialization within the host country?

■ The Agent-Based Identity Repertoire (ABIR) Model

One innovative way to study these phenomena is by creating virtual worlds. By abstractly specifying the tenets of constructivist theory and exploiting the calculation speed of computers, it becomes possible to overcome the difficulty of finding enough natural experiments in the real world to test important claims. By running the same world (or computer-created landscape) with randomly changing particular conditions (e.g., which individuals are activated on which identities at the beginning and exactly how these identities are distributed within the country), constant parametric conditions (e.g., how large the country is and how large an individual's identity repertoire is), and systematically varied experimental conditions (e.g., the openness of the world to outside pressures, the proportion of native identities displaced by immigrant identities, and the volatility of change in the incentives affecting expression of different identities), large numbers of "histories" of this world can be recorded.[2] The data contained in the record of these histories can then be analyzed for answers to questions about the effects of immigration or globalization on the dominance of particular identities within particular areas.

Before we describe some of the experiments we have conducted to address these questions and discuss our findings, it is necessary to briefly describe the ABIR model. Figure 4.1 shows two sets of sample ABIR screenshots. On the left are the sample 1 screenshots. Upper left is an ABIR landscape at time-step 0 (t = 0), and below it is the same landscape at t = 500. On the right side is sample 2. The same landscape at t = 0 again appears in the upper half of the figure (right side), but it now has a physical or territorial border surrounding a section of it. Below, on the right, is this landscape after 500 time-steps. Each history can be thought of as a sequence of 500 of these snapshots, with each snapshot representing a transformation of the previous snapshot according to rules matching the postulates of constructivist theory.

The landscape used to produce this history, and all the landscapes produced by this version of ABIR, are two-dimensional spaces inhabited by square-shaped agents. Each square-shaped agent interacts in each time period with agents in its "Moore neighborhood" (i.e., the eight agents that touch it on its sides and corners). Each agent at each time period is displayed in a particular color,[3] which represents the agent's currently activated identity. In addition, each agent is endowed, as are people in constructivist models, with a "subscription" of unactivated identities that complete its repertoire. The identities in each agent's repertoire, including its activated identity, comprise a subset of the total number of identities present in the repertoires of all agents in the landscape. Any identity, including those not initially present in a particular agent's repertoire, can, under the right set of circumstances, be activated by that agent or brought into its repertoire and then activated.

At the beginning of a run (which will be a history of the polity) the landscape can be "reseeded," randomizing both the distribution of activated identities and that of subscribed identities. In each period of history, or time-step, every agent observes the identities being activated by its eight neighboring agents (but not their unactivated identities—those we term *subscribed*) and the positive, negative, or neutral bias that the mass media are attributing to association with particular identities at

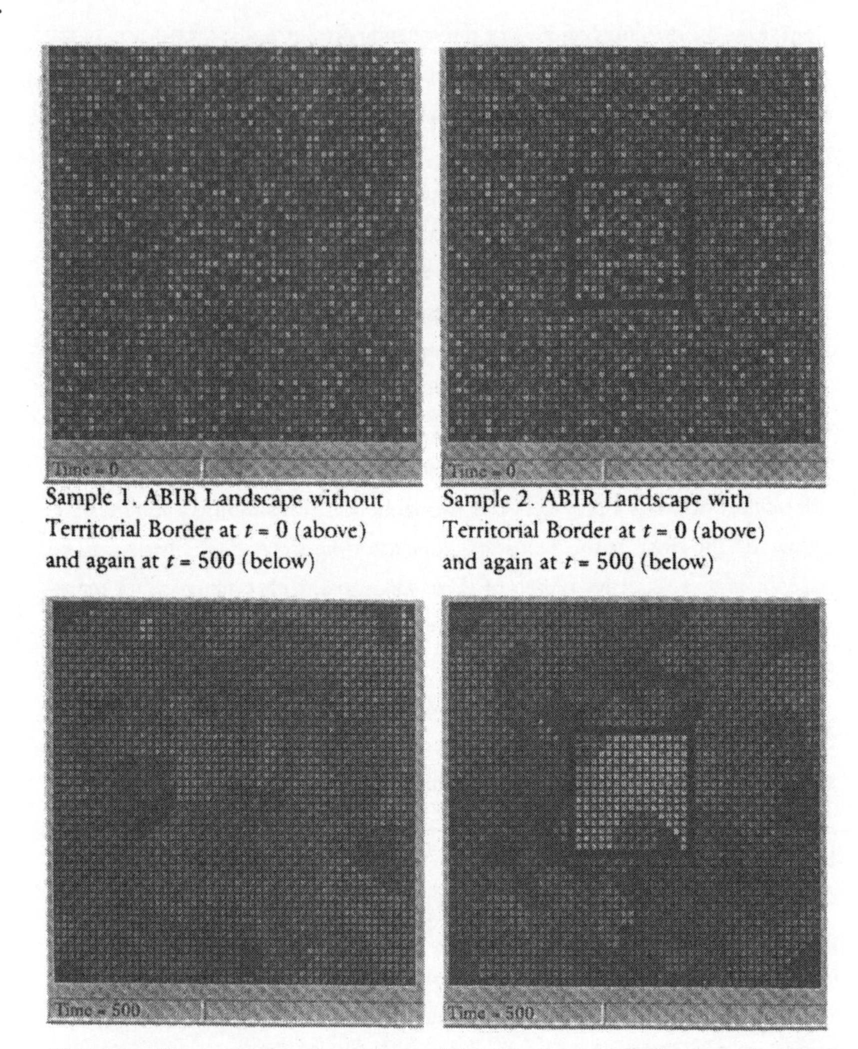

FIG. 4.1. Two screenshots from separate ABIR histories at beginning (t = 0) and end (t = 500) of run.

that time.[4] Each agent then responds to this information in one of three ways: (1) it maintains its currently activated identity into the next period, (2) it substitutes another identity from its repertoire as its newly activated identity, or (3) it replaces an identity within its repertoire with an identity previously absent from its repertoire. The simple algorithms that determine how and when an identity is included or extruded from an agent's repertoire or activated by an agent were designed to correspond as closely as possible to the assumptions of constructivist theory and to well-established findings in social psychology that, ceteris paribus, individuals have multiple identities, that different identities are activated under different circumstances, that those identity sets can change over time, and that patterns of activation are related to generalized perceptions of the value of those identities and to patterns of identity activation immediately accessible to an individual.[5]

In both this version and more advanced versions of ABIR, many characteristics of the agents and the spaces they inhabit that are relevant to constructivist theory can be varied—characteristics such as local influence level, resistance to new identities, size of the neighborhood experienced as local, speed of reaction to shifts in the incentives attached to activation on different identities, size of repertoire, spatial concentration of types of agents, turbulence in the rate and extent of change in identity bias signals, and so on. In the experiments reported here, however, most of these characteristics are held constant.[6]

■ Applying ABIR to Questions of Territory and Identity

In our exercise in virtual political geography, we imagine a hypothetical and vastly simplified territorial space in which one distinctive region, referred to as National Homeland, is a 15 × 15 box (225 agents) and World is the remainder of the 50 × 50 global landscape (2,275 agents). All residents of National Homeland are agents whose identity repertoires include exactly the same six identity options. These six identities (ways, as it were, of performing their national identity) are activated by approximately equal numbers of agents randomly distributed across the National Homeland territory. The inhabitants of World also have repertoires containing six identities (alien identities, from the perspective of National Homeland). These alien identities are drawn at random from the remainder of landscape identities not counted among the six National Homeland identities. Since all landscapes used in these experiments feature fifteen identities present in the global identity spectrum, this means that the six identities in the repertoire of each agent in World were drawn from a set of nine different identities. The nine identities are activated by approximately equal numbers of agents and randomly distributed across the territory designated World. It should be emphasized that we begin with the stark image that none of the World identities exist in National Homeland and that none of the National Homeland identities exist in World. From this basic template we then change distributions and locations of agents activated on different identities and containing different identity repertoires.

The abstract core of the theoretical issues of interest here can be framed as identifying the dynamics of the relationship between immigrants as carriers of naturalized identities, on the one hand, and native inhabitants of a territory who may be activated on different versions of a national identity but who all, at least initially, share the same set of identities in their repertoires, on the other. In particular we have focused on the impact of (1) the varying prevalence of immigrant identities activated or in the subscriptions of inhabitants of a country, (2) territorial/political borders versus mere identitarian/cultural borders, and (3) the presence or absence of a globalizing identity wave. These three independent variables were operationalized in the following ways.

The Prevalence of Immigrants and Naturalized Identities

An immigrant is operationalized as an agent within National Homeland whose activated identity at the beginning of the run ($t = 0$ of the history to be produced) is an identity formerly present only in World. As carriers of a new identity into National Homeland, immigrants are assumed to have affected the identity repertoire of every other inhabitant of National Homeland. Although only those activated on a non-native identity at $t = 0$ are deemed immigrants, that activated non-native identity is also placed in the subscription of each native inhabitant of National Homeland. To achieve statistical simplicity, and in support of our effort to focus on the abstract and fundamental dynamics of population composition and territory, the appearance of this immigrant or naturalized identity in each inhabitant's repertoire results in the removal of one specific previously available native identity from the repertoire of all nonimmigrant agents. Note that the native agents who are reclassified as immigrants in this reseeding process are those for whom an activated native identity is replaced by an immigrant activated identity.[7]

Let us consider an example that hopefully makes up in clarity what it may lack in realism. If the Dutch are deemed, as in our model, to have a total of six different Dutch identities in their collective repertoire and in the individual repertoire of each inhabitant of the Netherlands, then after the arrival of one wave of immigration, one-sixth of the agents within the Netherlands, or all those activated on one particular way of being Dutch (e.g., Catholic), are now activated on a specific immigrant identity instead. This naturalized identity (e.g., North African Muslim) is an identity previously present in the world that was not present in the Netherlands prior to the arrival of the immigrants. As noted, we simplify the immigrant-native society relationship by imagining a simultaneous replacement of, for example, the Catholic way of being Dutch within the nonactivated repertoires of all the other inhabitants of the Netherlands with the new non-native identity (Muslim).

To represent differing densities of immigrant populations or different degrees of cultural heterogeneity, three different versions of National Homeland have been constructed. In National Homeland One (NH1), one-sixth of the residents are immigrants. They are activated on one particular naturalized identity from World, replacing one native National Homeland identity. The unactivated identities in their repertoire include all five of the available native identities. The remaining five-sixths of the inhabitants of NH1 are natives activated in equal numbers on the five remaining native identities, but they now also have within their repertoires the naturalized identity activated by the immigrants. In National Homeland Two (NH2), two-sixths of the inhabitants are immigrants—one-sixth activated on each of two different World identities and all containing in their repertoires the four remaining native identities. That is to say, in NH2 two of the original native identities have been removed. The two-thirds of NH2 inhabitants comprising the native population are activated in equal numbers on the remaining four different native identities while containing within their repertoires each of the two non-native identities activated by the immigrants. Finally, in National Homeland Three (NH3), three-sixths of the residents are immigrants, activated in equal numbers on each of three naturalized World identities and containing within their repertoires all three available native identities. The other half of the inhabitants of NH3 are natives activated in equal parts on the remaining three native identities. Thus NH1, NH2, and NH3 differ incrementally from each other in the total number of immigrants and natives and in the number of different native and naturalized identities activated at $t = 0$, with NH1 containing the smallest percentage of immigrants (approximately 17%) and NH3 containing the largest percentage (50%).

■ Territorial/Political versus Identitarian/Cultural Borders

The second variable requiring operationalization is the distinction between physically enforced territorial borders on the one hand and identitarian borders on the other. We consider territorial borders as institutionalized demarcations of territory, usually associated with the officially sanctioned boundaries that separate one state from another. In contrast, identitarian or cultural borders, although they may correspond to territorial divisions, arise only from differences in the identities of clusters of inhabitants within a physically undivided space. For example, the territorial border of Albania is familiar and clear from map images of the Albanian state. But the Albanians outside that border who predominate in parts of Kosovo may present an identitarian or cultural border to non-Albanians that is different from physically enforced barriers to penetration by non-Albanians.

In the virtual worlds created by ABIR, the existence of a territorial boundary separating National Homeland from World is represented by a black border around National Homeland.[8] This barrier prevents inhabitants on one side from perceiving or being influenced by agents near them but on the other side of the border. Thus it prevents pressures against the border from being directly transmitted across it. The border does not affect the reception of identity bias signals—mass media reports that register uniformly in World and in National Homeland, whether surrounded by a boundary or not.

If this territorial border is removed, National Homeland agents along the frontier are exposed to the identities activated by World agents adjacent to them, and they are therefore subject to potential cascades of pressure to similarly activate on a non-native identity. In other words, the only barrier that remains in the frontier area is an identitarian boundary between National Homeland and World, a barrier whose resistance to the smooth transmission of patterns of change on one side into patterns of change on the other is produced only by the degree to which the patterns of identity activation and the composition of identity repertoires of National Homeland and World inhabitants differ.

■ The Presence or Absence of a Globalizing Identity Wave

Our third operationalized variable is one particular pattern of change that can arise within a landscape—namely, powerful but locationally specific cascades of change toward a single activated identity. Under the right set of conditions, a critical mass of agents activated on the same identity can emerge and spread outward in a cascade of cultural globalization. These conditions include all of the following: (1) a sufficient concentration of agents in World in a sufficiently small area activated on the same identity, (2) a sufficient number of other agents in that area with that identity within their repertoires, and (3) a significant boost in the bias attributed to that identity sustained for enough time periods. For our purposes we stipulate that the presence of a *globalizing identity wave* — from the point of view of National Homeland — is indicated in the history under review if, at the end of its 500 time-steps, at least one of the non-native identities of the initial National Homeland immigrants (i.e., at $t = 0$) is also an identity that is activated by at least 40 percent of all agents on the landscape (i.e., those inhabiting World and National Homeland combined). Thus, although the settings for bias volatility and the range of variation of these biases remain the same across all histories and all experimental conditions, cascades of globalization arise only in a subset of them. Furthermore, recall that it is only a smaller subset of these cascades that we define as globalizing identity waves — namely, cascades of National Homeland immigrant identities.

It is important to distinguish here between what we consider to be two distinct but related forms of globalization. When territorial boundaries are present between National Homeland and World, the only mechanism translating global events into changes in the complexion of National Homeland is the biases associated, globally, with different identities at different times. As noted, when territorial boundaries are present, all agents in National Homeland receive the same changing bias signals with the same volatility as do agents in World, but National Homeland inhabitants cannot be affected by cascades of a territorially concentrated and expanding globalizing identity. We call this *parametric globalization*, referring to the kinds of global changes that are relatively unaffected by state boundaries (e.g., climate, information transmission, world commodity prices, etc.). When territorial borders are removed, the effects of a globalizing identity wave can be felt directly and in a territorially specific fashion by inhabitants of National Homeland. Under these conditions, and when, in fact, a globalizing wave based on a World identity represented within National Homeland arises, we consider that *spatially focused globalization* (e.g., trade, immigration, tourism, direct investment, international communication, etc.).

■ ABIR Experimental Findings

We can now focus our attention on the findings of several of our ABIR experiments. In these experiments we examined two critical issues that help illuminate the interplay between spatially focused and parametric globalization and their impact on matters of territoriality and identity. First, the question of the long-term viability of activated native identities within National Homeland was explored. Second, the phenomenon of glocalization and its tendency to disrupt the balance among alternative native identities was investigated. Outcome measures at time-step 500 were computed for each of 600 histories for each of the three different levels of immigration represented by National Homeland One, National Homeland Two, and National Homeland Three (i.e., a total of 1,800 ABIR runs combined).

Two of our three key independent variables can be represented as the two dimensions of a two-by-two matrix as depicted in figure 4.2. The horizontal dimension designated across the top of the matrix distinguishes between whether, during any given history, National Homeland has a territorial border or instead only an identitarian border around its perimeter. The vertical dimension designated along the left side of the matrix distinguishes between whether a globalizing identity wave as defined earlier was present or absent during a given history. Within each of the four quadrants thereby created, we can use multiple sets of simulation runs with National Homelands to examine our third independent variable, namely, the proportion of immigrants present at time-step 0. Each ABIR history can be placed in one (and only one) of these quadrants. The most noteworthy features of each quadrant can be summarized as follows.

Quadrant I (Upper Left). This quadrant includes those histories characterized by a territorial border surrounding National Homeland and the absence of a globalizing identity wave. Of these experimental runs, we may say that National Homeland experienced weak parametric globalization and no spatially focused globalization. Here in Quadrant I, therefore, we would expect that the prospects for National Homeland to remain a region of distinctive identity and culture are most promising.

Quadrant II (Upper Right). In this second quadrant of the matrix are the histories produced when National Homeland had only an identitarian border and again, as in Quadrant I, when a globalizing identity wave was absent. Of these histories we may say that National Homeland experienced weak parametric globalization and weak spatially focused globalization. The designation of weak spatially focused globalization reflects the fact that although National Homeland was

Globalizing Identity Wave	Type of Border: Territorial	Type of Border: Identitarian
Absent	Quadrant I	Quadrant II
Present	Quadrant III	Quadrant IV

FIG. 4.2. Each ABIR history fits into one of the four quadrants of this 2 x 2 matrix, based upon the type of border surrounding National Homeland and the presence or absence of a globalizing identity wave.

not shielded by a territorial border from pressures emanating from World, these spatial pressures turned out not to meet our definition's threshold as a globalizing identity wave. It might be said, then, that Quadrant II contains those histories in which National Homeland happened to escape the potentially considerable impact on its native identities that might have been produced had the activated identities of its immigrants proved more appealing in World.

Quadrant III (Lower Left). In this quadrant are the histories produced when National Homeland had a territorial boundary and when a globalizing identity wave was present. Therefore, although National Homeland experienced no spatially focused globalization (by virtue of the territorial boundary), it was subject to strong parametric globalization. Under this combination of conditions, native inhabitants of National Homeland were fully protected from direct contact with outside influence. It may be expected, however, that National Homeland's native identities were still at risk because an immigrant activated identity was favorably advantaged in the landscape as a whole (by the high biases that tend to be associated with a globalizing identity wave).

Quadrant IV (Lower Right). In this final quadrant are the experimental histories produced when National Homeland was surrounded by only an identitarian border and when a globalizing identity wave developed. As a result, we may say that National Homeland experienced strong spatially focused and strong parametric globalization. This, then, is the quadrant of greatest apparent jeopardy for the preservation of native identities as the preferred ways of being within National Homeland. Not only is there no territorial border preventing World residents from impacting the identity choices of National Homeland inhabitants, but at least some of the immigrants within National Homeland have adopted an activated identity that has proved broadly popular beyond its borders. Snowball effects or cascades in favor of this identity can flow directly against National Homeland residents, reinforcing efforts by those activated on that advantaged identity to activate their neighbors.

The Long-Term Viability of Native Identities

The long-term viability of native National Homeland identities was assessed using two related outcome measures. The first measure was a percentage comparison of the absolute number of agents within National Homeland activated on any native identities at time-step 500 versus time-step 0. Figure 4.3 presents the average values for this variable in each quadrant for each of the three National Homelands (NH1, NH2, and NH3). Note that the number of ABIR histories varies for each condition (this is true for all subsequent results as well). These disparities in sample size reflect naturalistic differences in the occurrences of the circumstances that define each quadrant.

The findings reported in figure 4.3 for NH1, NH2, and NH3 are more readily interpretable against a baseline standard for a National Homeland in which there are no immigrants whatsoever at time-step 0. For this purpose, a separate set of ABIR histories (600 runs) was collected for a National Homeland Zero (NH0), in which all inhabitants were natives with repertoires that shared no identities with the residents of World. Half of these runs included a territorial border around National Homeland (i.e., no spatially focused globalization), and in all of these cases the number of agents activated on native identities necessarily remained constant at 100 percent over time because there was no possibility of an activated non-native identity appearing within the NH0 borders. For the other half of the runs,

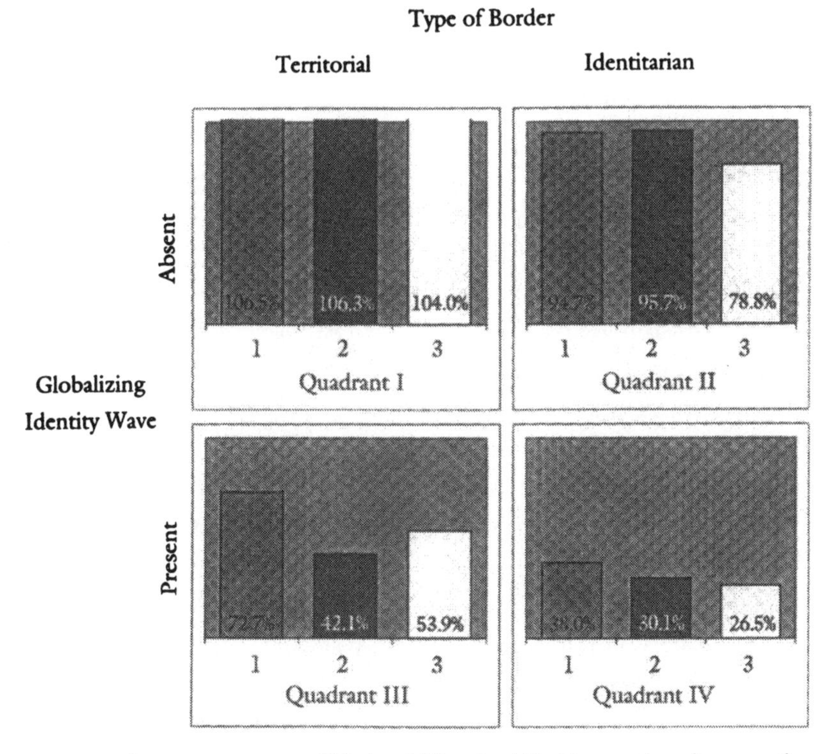

FIG. 4.3. Average percentage of National Homeland Residents activated on a native identity at t = 500 compared to t = 0. N's for NH1, NH2, and NH3 in each quadrant are as follows: Q1 (277, 255, 213), Q2 (268, 243, 216), Q3 (23, 45, 87), and Q4 (32, 57, 84).

NH0 and World were separated by an identitarian border only. In these cases, the average outcome measure value at t = 500 was 101.6 percent. This percentage indicates that when there were no immigrants within NH0 at t = 0 (and therefore by definition no possible globalizing identity wave), a territorial border was not needed to prevent switching from native to non-native activated identities.

In figure 4.3, a general but imperfect pattern is observable when comparing the three different National Homelands, NH1, NH2, and NH3. The percentage values for NH3—with the greatest number of immigrants initially—were consistently lower than the corresponding values for NH1. Thus, the larger initial presence of immigrants in NH3 appeared to apply more pressure on the native population of National Homeland to convert to an activated non-native identity. But beyond this finding, there are also noteworthy differences in the magnitude of this effect depending upon which of the four quadrants of our matrix we consider.

For example, on average the impact of immigrants on native identity representation over time was nonexistent when National Homeland was insulated by a territorial border and none of the immigrant identities within National Homeland established a globalizing identity wave (i.e., Quadrant I). Indeed, under these conditions of no spatially focused globalization and only weak parametric globalization, there was a slight tendency for the proportion of National Homeland residents activated on native identities to actually increase, rather than decrease, over time. In other words, rather than fighting a losing battle to maintain their representation within National Homeland, inhabitants expressing native identities not only held their own but actually succeeded in convincing immigrants to activate on native identities as well. In Quadrant II the pairing of weak spatially focused and weak parametric globalization produced a decline over time in the percentage of National Homeland residents activated on a native identity—more so when immigrants were more highly represented at $t = 0$ as in NH3—but these losses were still relatively modest when compared with the effects witnessed in the matrix's lower quadrants.

Quadrants III and IV have in common the presence of a globalizing identity wave—that is, the histories included in both quadrants featured conditions of strong parametric globalization. They differ from each other in regard to whether it was a territorial border (Quadrant III) or only an identitarian border (Quadrant IV) that served to protect National Homeland's perimeter from the influence of neighboring World residents. As the average percentages in Quadrant III indicate, when confronted with a globalizing identity wave, even National Homeland's impermeable territorial border was clearly insufficient to prevent widespread conversion to activated immigrant identities by National Homeland natives. The magnitude of this conversion was even more pronounced in Quadrant IV, where exposure to strong spatially focused globalization as well as strong parametric globalization produced dramatic declines in National Homeland residents activated on native identities at $t = 500$. Still, the protective role that territorial boundaries do play is clearly reflected in the fact that the average prevalence of native identities at $t = 500$, across all homeland conditions, is 78.7 percent higher in Quadrant III than in Quadrant IV.

The second outcome measure used to assess the long-term viability of native

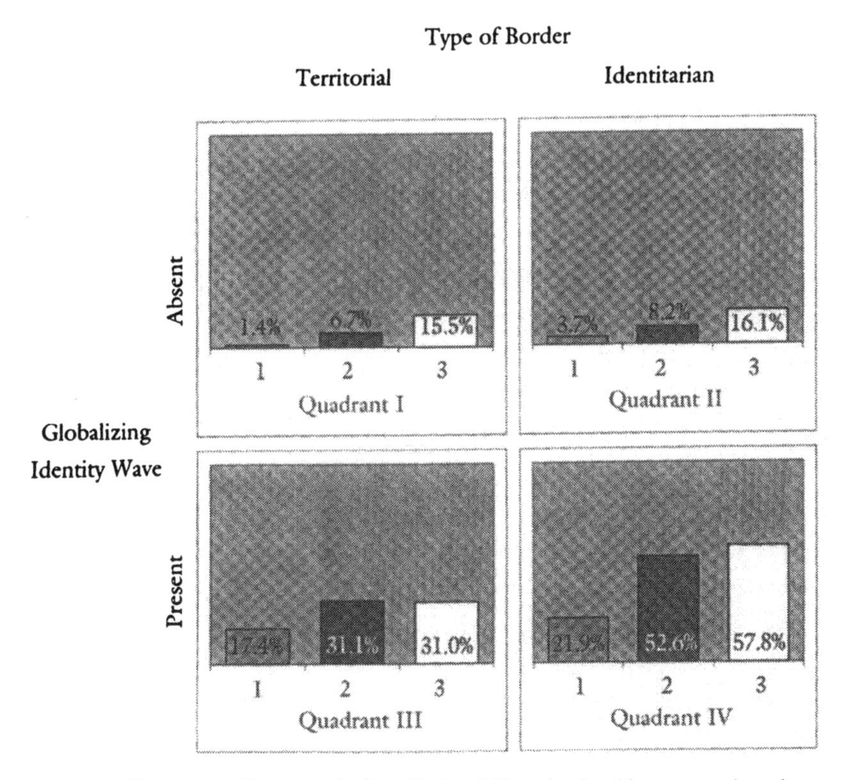

FIG. 4.4. Percentage of runs in which *no* National Homeland residents are activated on a native identity at $t = 500$. *N*'s for NH1, NH2, and NH3 in each quadrant are as follows: Q1 (277, 255, 213), Q2 (268, 243, 216), Q3 (23, 45, 87), and Q4 (32, 57, 84).

National Homeland identities was a simple count of the number of ABIR histories in which at $t = 500$ there were no longer any National Homeland inhabitants activated on any native identities. In other words, we asked the question, How often does National Homeland become the exclusive province of residents activated on non-native identities? This eventuality occurs only when all of the native inhabitants of National Homeland are persuaded to switch to one of the identities brought into National Homeland by the immigrants from World. Note that there were no cases of extinction in baseline runs where there were no immigrants in National Homeland at $t = 0$. Figure 4.4 presents results in the form of the percentage of histories—*extinction histories*—in which total activated native identity extinction occurs. Again, the quadrants of our matrix reveal noteworthy differences.

In Quadrant I, National Homeland's native population was doubly protected by an impenetrable territorial border and the failure of a globalizing identity wave to emerge (i.e., no spatially focused globalization and only weak parametric globalization). Under these conditions extinction histories represented a very low percentage of the overall runs. They ranged from a low of 1.4 percent for NH1 up to a modest 15.5 percent for NH3. Comparable levels of extinction were found in Quadrant II (weak spatially focused globalization coupled with weak parametric globalization), indicating that by this measure the presence or absence of a territorial (compared to a merely identitarian) border was inconsequential as long as National Homeland was not subjected to an alien globalizing identity wave.

If we now move to the lower half of the matrix, we see substantially higher rates of activated native identity extinction. In Quadrant III, despite the presence of a territorial border surrounding National Homeland (i.e., no spatially focused globalization), the percentage of histories in which all residents relinquished their native activated identities in favor of non-native activated identities was elevated substantially above the corresponding figures for Quadrant I for all three National Homelands. Thus, we can see that the presence of a globalizing identity wave (i.e., strong parametric globalization) had a significant impact.

The results were even more striking for Quadrant IV, where National Homeland's boundary was only identitarian in nature and where a globalizing identity wave occurred. The extinction rates were sharply higher for Quadrant IV than Quadrant III, indicating that a territorial border eliminating the influence of spatially focused globalization had a meaningful protective role when strong parametric globalization pressures were present. Indeed, in NH3, where natives and immigrants resided in equal numbers at $t = 0$, for well over half of the histories in Quadrant IV there were no longer any inhabitants activated on a native identity at $t = 500$. Also, the greater disparities between Quadrants IV and II than between Quadrants III and I reveal that the presence or absence of a globalizing identity wave made much more difference when the barrier between National Homeland and World was solely identitarian rather than when it was territorial.

The Parochialization of Native Identities

Thus far we have focused on ABIR findings pertaining to the protection of native expressions of distinctive national identities in a country moving toward greater commonality and association with other polities. We now shift our attention to reactive nationalism or, more generally, ***glocalization,*** a phenomenon equally of interest to this discussion because it raises questions about the preservation of diversity within an individual country in regard to the breadth of variety in the manner in which native national identity is expressed. The term itself refers to a syndrome of parochial mobilization arising within states exposed to severe globalizing pressures. These manifestations of what Barber has iconized as "Jihad" in reaction to "McWorld" press the nation-state from within and below even as it is being squeezed from above and without by global economic and cultural forces. We make no claims here as to the potency of this kind of threat to the integrity of the nation-state. But by defining glocalization in terms of a parochialization of identity—a reduction in the manifest diversity of native ways of performing the national—we can examine the extent to which glocalization can arise as a typical result of certain kinds of encounters in a constructivist world rather than as a pathology associated with particular cultures or civilizational levels.[9]

Our ABIR experiments measured parochialization effects in two different but related ways. The first measure evaluated native identity concentration in National Homeland at time-step 500 for each run by using the Herfindahl Index, which is traditionally employed to describe the extent of concentration in an industry or market for econometric or antitrust purposes. In our case, this index was the sum of the squares of the proportional representation of each native identity (relative to the total number of National Homeland agents activated on native identities). Since ABIR histories in which all native identities were extinguished could not be used to assess the concentration of remaining native identities, these runs were excluded from the index calculations.

Figure 4.5 presents these average Herfindahl Index values for each quadrant. The highest possible score, reflecting maximum concentration, is 1.00, indicating that all remaining agents activated on a native identity were in fact activated on the same identity. The minimum concentration scores varied from one National Homeland to another, based on the number of available native identities at $t = 0$. These minimum possible Herfindahl Index values were .20, .25, and .33 for NH1, NH2, and NH3, respectively. It should also be noted, for comparison purposes, that the index score at $t = 500$ for 600 baseline histories in which no immigrant identities were placed in National Homeland (i.e., NH0) was .61 (with a minimum possible value of .17). This value in itself indicates that sub-

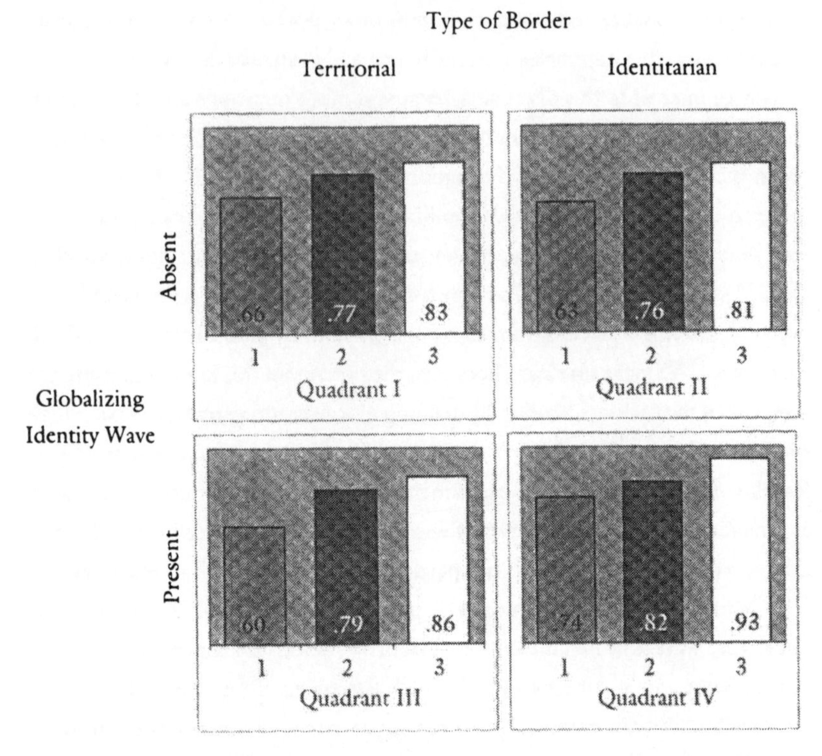

FIG. 4.5. Average Herfindahl Index as measure of concentration of native identities among National Homeland residents at t = 500 for histories in which complete extinction did not occur. *N*'s for NH1, NH2, and NH3 in each quadrant are as follows: Q1 (273, 238, 180), Q2 (258, 223, 181), Q3 (19, 31, 60), and Q4 (25, 26, 36).

stantial consolidation among native identities occurred even without outside influence.

The findings here reveal an important general trend, with all four quadrants displaying the same pattern: at t = 500 there are, on average, increasing levels of concentration among remaining activated native identities as we move from NH1 to NH2 to NH3. The average Herfindahl Index scores across all quadrants were .66 for NH1, .79 for NH2, and .86 for NH3. That is, the presence of more immigrants at t = 0 (and therefore fewer native residents and fewer native identities in the repertoire) tended to produce greater concentrations of activated native identities over time, regardless of border type or presence versus absence of a globalizing identity wave. However, beyond this overall pattern, we see a noteworthy accentuation in this consolidation process in Quadrant IV. Here, where the native National Homeland agents were exposed to strong spatially focused globalization (the border between National Homeland and World is strictly identitarian) and a globalizing identity wave (i.e., strong parametric globalization), the average Herfindahl Index scores reached higher levels than in the other three quadrants. Indeed the .93 value for NH3 in Quadrant IV is quite close to the maximum possible value of 1.00.

A related second outcome measure of parochialization sheds further light on the constriction in the range of alternative activated native National Homeland identities over time. The variable we employed for each history was the number of different native National Homeland identities still activated (by at least one agent) at $t = 500$. Individual runs in which there were no activated native identities left at $t = 500$ were excluded from the analysis, again because such histories could not be used to measure changes in native identity concentration over time. This means that the minimum possible value for the outcome measure was 1—rather than 0—since at least one agent activated on a native identity was required for inclusion.

Note that for NH1 there were initially 5 different activated native identities, for NH2 there were initially 4 different activated native identities, and for NH3 there were initially 3 different activated native identities. Figure 4.6 presents the average number of activated native identities remaining at $t = 500$ for each quadrant of the matrix. In considering these values, it is again important to compare them with baseline figures established from histories in which National Homeland had no immigrants whatsoever at $t = 0$ (i.e., for NH0, in which there were initially 6 different activated native identities). For these 600 runs the average number of different native identities remaining at $t = 500$ was 2.92.

Figure 4.6 demonstrates a general trend across quadrants for this variable similar to that observed for the Herfindahl Index scores in figure 4.5. All of the average values were well below the corresponding original complement of native identities for each National Homeland, indicating a glocalization-relevant effect of substantial declines in activated native identity diversity over time. In addition, within each quadrant the average number of different activated native identities at $t = 500$ was proportionally smaller when more immigrant agents, and fewer native agents, were included in National Homeland at $t = 0$.

Not surprisingly, the most dramatic decline in the diversity of activated na-

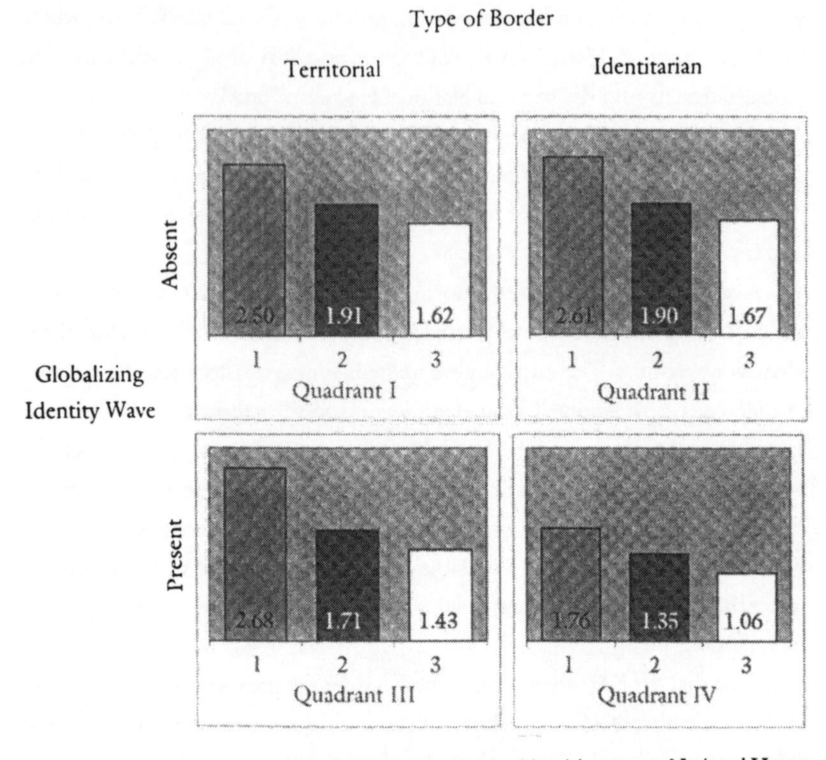

FIG. 4.6. Average number of different activated native identities among National Homeland residents at t = 500 for histories in which complete extinction did not occur. N's for NH1, NH2, and NH3 in each quadrant are as follows: Q1 (273, 238, 180), Q2 (258, 223, 181), Q3 (19, 31, 60), and Q4 (25, 26, 36).

tive identities occurred in Quadrant IV, where the combination of strong parametric globalization and strong spatially focused globalization took a severe toll. For example, the 1.06 value for NH3 is barely above the minimum possible value of 1.00 (which equates with a complete loss of diversity). This value also represents an average loss of 1.94 (activated native identities) over time, which is very close to the maximum possible loss of 2.00 (i.e., from the original 3 activated native identities in NH3 to only 1 identity). On a percentage basis, this average decline is fully 97 percent of the largest decline possible. Also noteworthy is the greater disparity between Quadrants IV and II than between Quadrants III and I. This distinction again indicates that the presence of a territorial border around

National Homeland served to limit, to some degree, the impact of a globalizing identity wave.

■ Conclusion: Territorial Borders, Globalization, and the Parochialization of Native Identities

Formal models, whether statistical, game-theoretic, or, as in our work, agent-based, are intended to identify the importance of variables otherwise obscured or underappreciated. The ruthlessness of their abstractions means that deploying these models drains enormous amounts of detail from the situations being modeled. Thus the successful use of formal models cannot be measured by the production of a recognizable and satisfying snapshot of the composite set of circumstances and forces under investigation. Instead, the usefulness of any of these approaches is that by teaching us something about unabstracted realities, they can show the potency of the variables they do focus on even if most of the actual content is removed. This is, of course, as it should be. In physics, abstract models filled with frictionless planes, constant velocities, and perfect spheres illuminate the fundamental importance of key relationships between mass, motion, and acceleration. Such models teach us about automobile crashes, even though all of what we know, or can know, about speed, location, malfunction, gasoline, rust, human error, irregularity of materials, and road conditions is removed from the target of analysis.

In the experiments reported here, we have indeed ruthlessly drained cultural distinctiveness, history, and many other elements of interest from the problem of studying territorial borders and their relationship to the effects of mixing new identities into a national polity. However, we have strived to organize our research so as not to contradict what is generally accepted—and believed by us—to represent the fundamentally constructivist dynamics of identity formation and change. While we have used Europe as a source of folk examples and as a loose narrative for explaining the strategic intent of our experiments, we have made no claims to be modeling anything specific about territory, identities, or processes of migration and cultural transmission in contemporary Europe.

Yet we believe we have learned some things that may help us and others better understand the forces operating in Europe. One conclusion we can draw is that, indeed, territory does matter, but it matters under some circumstances more than others. When globalizing forces are relatively weak or nonexistent—when the outcome of processes of identity change are determined more or less completely by the resolution of forces operating within the state—a territorial border that physically demarcates the separation of the state from the rest of the world is largely unnecessary for the protection of the survival and predominance of native identities within the polity. This is the case especially when the proportion of inhabitants embracing naturalized, formerly alien identities is relatively small.

Thus, noting the NH3 bar in Quadrant II of figure 4.3, we see that even in a polity where at $t = 0$ half of the residents are activated on immigrant rather than native identities, an average of 78.8 percent of inhabitants initially expressing themselves in native idioms are still doing so at the end of the history under observation. In NH1 and NH2, where the proportions of immigrant identities are lower, the corresponding averages at $t = 500$ are well over 90 percent. In other words, under these conditions identitarian borders—distinctions between the polity and the world arising out of the complexion of identity commitments rather than out of physical or territorial borders per se—can provide reassurance to inhabitants embracing native variants of the national identity that they will not be submerged or made irrelevant by the presence of naturalized, previously alien identities.

However, under conditions of strong parametric and spatially focused globalization (Quadrant IV) and substantial cultural heterogeneity (NH3), territorial boundaries play a crucial role in preventing the presence of naturalized identities, and inhabitants expressing those identities, from swamping the state and reducing or even displacing native ways of performing the national. Again with reference to figure 4.3, we note that native identity survival rates are significantly reduced by the presence of a globalizing identity wave (Quadrants III and IV compared to Quadrants I and II) but, just as dramatically, that the presence of a territorial border (Quadrant III) affords substantially more protection against the impact of globalizing cascades of identity transformation than does the presence of an identitarian border only (Quadrant IV).

The same conclusion emerges from consideration of the data displayed in figure 4.4. Here we compare rates of extinction of all activated native identities in each of the three National Homelands under each of the four conditions of exposure to parametric and spatially focused globalizing forces. Again we find that when globalizing forces are relatively weak (Quadrants I and II), at least some inhabitants of each National Homeland remain activated on a native identity at t = 500 in a very high percentage of the histories. Unsurprisingly, we also observe that as the initial presence of non-native identities increases (moving from NH1 to NH2 to NH3), the number of extinction histories also rises (but still remains small). We can see as well that, under these conditions of weak globalization pressures, there is little difference between territorial and identitarian borders. However, comparing Quadrants III and IV to Quadrants I and II reveals that potent globalizing forces do substantially increase the extinction rates of activated native identities. Indeed the lowest extinction rate when a globalizing identity wave was present (17.4% for the native identity-rich and territorially bounded NH1 in Quadrant III) was greater than the highest extinction rate when such a cascade was absent (16.1% for the immigrant-rich and identitarian-bordered NH3 in Quadrant II).

By comparing Quadrant III, where a territorial border is present, to Quadrant IV, where only an identitarian border is present, we can also see the effect of strong spatially focused globalization in concert with powerful parametric globalization. The rates of extinction for activated native identities are markedly higher in Quadrant IV than in Quadrant III, where territorial borders block the impact of spatially specific cascades of identity change emanating from beyond National Homeland. This impact of strong globalizing forces on the likelihood of activated native identity extinction is especially striking when the proportion of immigrants in National Homeland rises from one-sixth of the population (NH1) to one-third and one-half (NH2 and NH3). An apparent threshold effect is manifested in the particularly sharp increase in extinction histories under conditions where spatially focused globalization is added to parametric globalization. In Quadrant IV we see the incidence of native identity extinction rising from 21.9 percent in NH1 to over 50 percent in both NH2 and NH3.

The other significant finding to emerge from our experiments pertains to glocalization, or reactive nationalism, a phenomenon that figures in many efforts to understand the impact of migration on different European countries. Switzerland is the European country with the highest proportion of foreign-born inhabitants — 20 percent. Although in some parts of Switzerland the proportion of foreign-born reaches 30 percent, and although these inhabitants do not (yet) enjoy Swiss citizenship, we can see that our three different National Homelands, featuring proportions of one-sixth, one-third, and one-half naturalized inhabitants hailing from outside National Homeland borders, are on the high end of what most European countries have yet experienced. Nonetheless, if we take into account waves of immigrants who have become naturalized citizens of different European countries, or whose children and grandchildren are raised as nationals of their countries of habitation, then one can appreciate that the proportions used in our experiments are not terribly discrepant with European realities. Considering the invigoration of anti-immigrant movements in Britain, France, Germany, Denmark, and elsewhere and the concomitant rise of xenophobic, nativist nationalisms, the questions our experiments pose in the abstract are of direct relevance to the interactive meaning of territory, population, and boundaries in contemporary Europe.

Since our model is thoroughly abstracted from any particular country, we may say that our findings in regard to nativist parochialization in response to globalization and immigration are independent of any specific cultural or psychological disposition seen as peculiar to particular European countries. We also note that the patterns of parochialization that we find indicate a narrowing of native mobilization but not the direction that narrowing will take (reaction against globalization or bandwagoning mobilization in imitation of it). Thus our work cannot be read as a guide to understanding political culture change in any particular European country or as a prediction of which strains in the native identity repertoire will prosper at the expense of others. However our study of the dynamics of processes leading to nativist parochialization can suggest (1) whether such developments are to be expected in any polity whose borders open wide to global market forces and the movements of people and ideas from within and from outside Europe and (2) whether there are trigger or tipping points beyond which dramatic and frightening challenges to native inhabitants are especially likely.

We did find a glocalization effect, but it was not as generalized as the effects we found globalization and cultural mixing to have on the prevalence of native identities. Instead we observed a tendency, under all experimental conditions, including those with weak and with strong spatially focused and parametric globalization, for increasing levels of cultural mixing (moving from NH1 to NH2 to NH3) to produce moderately more parochialization among activated native identities as measured by the Herfindahl Index. For example, the National Homeland index averages (see fig. 4.5) are remarkably similar for Quadrants I, II, and III—where spatially focused globalization forces were either weak or absent—and in most conditions well above the baseline Herfindahl Index value of .61 (for those NH0 histories in which National Homeland had no immigrants at all). However, in Quadrant IV, where strong spatially focused globalization and strong parametric globalization pressures were both present, we found evidence (Herfindahl Index averages ranging from .74 for NH1 all the way up to .93 for NH3) of the kind of intensive parochialization associated with the glocalization or reactive nationalism phenomenon. In sum, then, we observed a tendency toward convergence toward a subset of national identities (even in the absence of strong globalization pressures), a significant increase in the tendency toward that convergence when cultural mixing is introduced, and an even greater increase in this parochialization tendency when the absence of territorial barriers directly exposes residents of the polity to the effects of globalizing identity waves.

The data displayed in figure 4.6 provide another cut at the same phenomenon. Here patterns of parochialization of activated native identities appearing in our histories are examined by measuring the average number of different native identities still being expressed by inhabitants at the end of the measured history, in each National Homeland and under each set of globalizing pressures. The lower the number of different activated native identities, the higher the degree of parochialization. Again we find under all experimental conditions, regardless of level of spatially focused or parametric globalization, that increasing levels of cultural mixing (moving from NH1 to NH2 to NH3) tend to produce more parochialization among activated native identities. As measured in this fashion, it appears that the most marked change in parochialization occurs in the transition between NH1 and NH2 (i.e., between circumstances of low to medium presence of immigrants). Finally, as we noticed when using the Herfindahl Index, we again see here a significant increase in parochialization when the polity, however mixed culturally, is deprived of a territorial boundary to protect it from the direct effects of strong spatially focused globalization (i.e., Quadrant IV).

That we have documented certain effects and relationships in the virtual worlds we have created with ABIR does not prove that these patterns exist in the real world. Nor does our work show that if they do exist, they are produced by the algorithmic processes and mechanisms of emergence identified in our analysis of the virtual world. The real world, after all, is filled with specificities, qualifications, and gradients absent from our analysis. Many of these details suggest avenues for further research with ABIR, now that we have shown ourselves able to produce virtual polities whose behavior—in terms of identity change, responses to different kinds of globalization, and responses to different proportions of cultural mixing—seem to correspond to some real-world patterns.

For example, boundaries in the real world are neither entirely sealed nor entirely open, and so in other experiments we are examining the effects of gradually increasing the porosity of territorial borders in relation to globalizing pressures. In the real world the arrival of immigrants or identities does not appear as a perfect substitution of one new identity for one previously present identity at both the level of activation and that of subscription. In the real world inhabitants of states have different sizes of repertoires and are differentiated in other ways—influence over one another, alertness to change in the environment, flexibility in response to opportunities for change, and so on. Each of these variables can be captured in the virtual world using ABIR, but in this study we have followed the KISS ("Keep it simple, stupid") strategy advocated by a leading developer of agent-based modeling techniques in the social sciences, Robert Axelrod (1997, 5). In this way we can systematically evaluate contributions to our understanding of the phenomena of interest as more complexity is added to the model.

It is this strategy that allows us to say some significant things, even though our virtual operationalizations are so brutally simplified. We can fairly conclude, for example, that the fundamental assumptions of constructivist identity theory are perfectly compatible with patterns of identity change and mobilization associated with globalization and glocalization in Europe; that globalizing forces acting in a localized fashion across borders must be distinguished from those whose effects are less spatial and more parametric; that emotional, socioeconomic, or historical variables, although they may add to our understanding of the patterns we have traced, are not necessary for explaining why they arise; and that territory and territorial boundaries matter, especially when significant cultural mixing has already occurred and when globalizing pressures are intense.

NOTES

This research was supported by grants from the Carnegie Corporation of New York and the Solomon Asch Center for Study of Ethnopolitical Conflict at the University of Pennsylvania. Comments for the authors may be directed to royeidel@psych.upenn.edu and ilustick@sas.upenn.edu.

1. Concerning the trend in immigration research toward the application of constructivist theories of nationalism, ethnicity, and identity, see Foner, Rumbaut, and Gold 2001.

2. From the perspective of $t = 0$ (before the beginning of a simulation), the run can be considered a future. From the perspective of the end of the simulation, it can be considered a history.

3. For full-color images of the figures in this chapter, see www.polisci.upenn.edu/abir/_private/publications/EideLust.doc.

4. This bias appears in the model as a negative or positive integer or as zero. When added to the assessment of the different identities appearing in an agent's neighborhood, the bias value affects the behavior of the agent as regards maintaining the agent's activated identity, changing it, or substituting a new identity for a currently subscribed one. Bias values change randomly within a specified range to represent variation in the attractiveness of various identities. In the experiments reported, the bias assigned to a particular identity in a particular time-step could change with a probability of .0009 to take on integer values between −2 and +2 inclusive.

5. For a complete specification of these algorithms, or "micro-rules," see Lustick and Miodownik 2000. For useful illustrative treatments of social identity theory as a prominent approach to identity issues in social psychology based on constructivist assumptions, see Abrams and Hogg 1990 and Hogg and McGarty 1990. See also Shin, Freda, and Yi 1999.

6. For published work reporting results of research conducted so far, see Lustick 2000; jasss.soc.surrey.ac.uk/3/1; Lustick and Miodownik, 2000; and Lustick and Miodownik 2002. To download an executable version of the ABIR program, along with a manual explaining its use, see www.psych.upenn.edu/sacsec/abir/_private/Simulation%20Software.htm. See also "PS-i," a powerful toolkit for producing models similar to ABIR but capable of much more refinement (see Lustick 2002).

7. In the histories displayed in figure 4.1, the National Homelands contain four different activated native identities (particular shades of red, green, blue, and khaki) and two different activated immigrant or naturalized identities (particular shades of gray and brown).

8. Return to figure 4.1 to note the operationalization of the territorial boundary in the history displayed on the right side of the figure.

9. For more detailed discussions of glocalization effects, see Axford 1995; Barber 1995; and Ram 1998.

FIVE

State Formation and Supranationalism in Europe

The Case of the Holy Roman Empire of the German Nation

ROLAND AXTMANN

The State in Europe and the Changing European Space

In *The Rise and Decline of the State,* Martin van Creveld (1999) reminds his readers of the fundamental distinction between *government* and *the state.* "The former," says van Creveld, "is a person or group which makes peace, wages war, enacts laws, exercises justice, raises revenues, determines the currency, and looks after internal security on behalf of society as a whole, all the while attempting to provide a focus for people's loyalty and, perhaps, a modicum of welfare as well" (415). The state should be understood as a corporation, hence as possessing a legal persona of its own: "[I]t has rights and duties and may engage in various activities *as if* it were a real, flesh-and-blood, living individual. The points where the state differs from other corporations are, first, the fact that it authorizes them all but is itself authorized (recognized) solely by others of its kind; secondly, that certain functions (known collectively as the attributes of sovereignty) are reserved for it alone; and, thirdly, that it exercises those functions over a certain territory inside which its jurisdiction is both exclusive and all-embracing" (1). In his book, van Creveld is at pains to emphasize that the state is merely one of the forms the organization of government has assumed.

This fundamental distinction between government and the state has long been forgotten in mainstream political and social science. The revival of historical macrosociology, however, has issued in a renewed interest in the concatenation of those circumstances that welded government and the state together. The description and explanation of the transformation of the multiplicity of overlapping and divided authority structures of the medieval polity into the modern, territorially consolidated sovereign state has been a central concern in historical sociology. Arguably, it is the current discussion of globalization, and, in particular, the (alleged) decline of the power and capacity of the modern state in the age of globalization, that raises the question as to how government and the state may come apart again.

In retracing the history of this development, a number of themes have been pursued. In premodern Europe, political authority was shared among a wide variety of secular and religious institutions and individuals—kings, princes and the nobility, bishops, abbots and the papacy, guilds and cities, agrarian landlords and bourgeois merchants and artisans. The modern state project aimed at replacing these overlapping and often contentious jurisdictions through the institutions of a centralized state. This endeavor was legitimated by the theory of state sovereignty, which claimed the supremacy of the government of any state over the people, resources, and, ultimately, all other authorities within the territory it controlled. State sovereignty meant that final authority within the political community lay with the state, whose will legally, and rightfully, commanded without being commanded by others, and whose will was thus absolute because it was not accountable to anyone but itself.

On the one hand, the sovereignty of the state presupposed the eradication of internal contenders for supremacy. On the other hand, the idea of sovereignty was premised on the notion of unity. In Thomas Hobbes's classical formulation, sovereignty means the reduction of all individual wills "unto one Will," thus establishing "a reall Unitie of them all, in one and the same Person," and "the Multitude so united in one Person, is called a Common-Wealth, in latine CIVITAS" ([1651] 1991, 120). The sovereign is the "very able Architect" who designs "one firme and lasting edifice" by abolishing diversity and irregularities or by explicitly sanctioning them (221): "Where there be divers Provinces, within the Dominion of a Common-Wealth, and in those Provinces diversity of Lawes . . . [they] are now Lawes, not by vertue of the Praescription of time, but by the Constitutions of their present Soveraigns" (186).

As a result of historical developments that span several centuries, governing in the modern sovereign state took on the form of the artful combination of space, people, and resources in territorialized containments, and the policing, monitoring, and disciplining of the population within these spaces became the foundation, and the manifestation, of state sovereignty. The modern territorial state came into existence as a differentiated ensemble of governmental institutions, offices, and personnel that exercises the power of authoritative political rule-making within a continuous territory that has a clear, internationally recognized boundary. It thus possesses *internal* sovereignty that is typically backed up by organized forces of violence and that grounds the state's *external* sovereignty vis-à-vis other states and its demands for noninterference in internal matters. Hence, sovereignty has a spatial dimension in that it is premised on the occupation and possession of territory. The spatial dimension manifests itself most clearly in the drawing of territorial boundaries that separate the inside from the outside. This territorial exclusion is, in turn, the prerequisite for identifying the source of sovereignty within the bounded territory and for defining "us" in contradistinction to "them."

Historical sociologists differ in their explanations of the formation of the modern state (Axtmann 1993). Yet, there is widespread agreement that government, in order to be effective and efficient, had to be organized in the modern state. Many paths may lead to modern statehood, but ultimately, "states proved more effective than cities or city-leagues in either waging war, trading with other units, keeping order within their realm, or extracting resources from its subjects—or a combination of these factors" (Jönsson, Tägil, and Törnqvist 2000, 68). Charles Tilly (1990) has emphasized most strongly both the different trajectories to modern statehood and the ultimate convergence on one model. Tilly argues that states as geopolitical actors depend on economic resources provided by dominant classes; that the most powerful states in military terms depend particularly on capital resources and thus on capitalist development; and that the closer the cooperation between states and capitalists, the greater the ease with which resources can be extracted and mobilized. He avers that "[t]he various combinations of coercion and capital across the European map shows us multiple paths of state formation and an ultimate convergence on states with high concentrations of both capital and coercion" (Tilly 1994, 23). The capacity to monitor, contain, seize, and redistribute resources within national boundaries has been an important aspect of state-building and state power. These resources included "not only goods and money but also land, natural resources, labor, technology, capital, and information" (26). Arguably, this capacity is currently declining, and after several centuries in which capital and coercion converged under state command, they now appear to be separating. For Tilly, this raises the question whether we are witnessing the disintegration of consolidated states.

At least in Europe in the past, the state was considered to be the ultimate power that could impose, and enforce, order within a territory. Political rule in general, and the regulatory, steering, and coordinating capacities of the state in particular, has been territorially bounded in its reach. The success of the nation-state in the last two hundred years or so, as well as its universality and legitimacy, were premised on its claim to be able to guarantee the economic well-being, the physical security, and the cultural identity of the people who constitute its citizens. However, ever more societal interactions cross borders and become transnational. They become therefore detached from a particular territory. The links between the citizens and the nation-state are becoming ever more problematic. The citizens demand political representation, physical protection, economic security, and cultural certainty. But in a global system, made up of states, regions, international and supranational organizations, nongovernmental organizations, and transnational corporations, the nation-state finds it increasingly difficult to accommodate these interests and mediate between its citizens and the rest of the world. Arguably, we are witnessing a steady replacement of the centralization and hierarchization of power within states and through states in the international system by the pluralization of power among political, economic, cultural, and social actors, groups, and communities within states, between states, and across states.

It is within this general *problematique* of the restructuring of the architecture of the modern polity that, ever more frequently, analyses of European integration are being placed. Europe has developed a particularly high degree of regional cohesiveness with a complex governance structure now emerging within the European Union. How best to categorize this emerging EU structure is a heatedly contested question. Joseph Weiler provides a crisp summary (1999a, 272–78). Some analysts see states as the key players and governments (primarily the executive branch) as the principal actors and privileged power holders. The Union is the international arena in which governments interact and, in pursuit of their respective interests, negotiate and bargain with each other. Other analysts agree that states are privileged players but perceive the Community/Union to be more than simply an arena in which international diplomacy is conducted. But contrary to the view of the *intergovernmentalists,* these *supranationalists* assign great importance to Community institutions such as the Commission, the Council, the European Parliament, and the European Court of Justice: they, too, are important players in the European polity. Finally, there are those analysts who take an *infranational* approach. According to Weiler, infranationalism is based "on the realization that increasingly large sectors of Community norm creation are done at a meso-level of governance." The actors involved tend to be, not governments or the Community institutions, but rather, "both at Union and Member state levels, administrations, departments, private and public associations, and certain, mainly corporate, interest groups" (1999a, 98, 273).

Weiler rightly suggests that governance in the European Union is a mixture of intergovernmentalism, supranationalism, and infranationalism. This messy structure of political regulation, coordination, and decision making is likely to be further complicated as a result of differentiated integration, instigated by the 1992 Treaty on European Union and the Treaty of Amsterdam (Neunreither and Wiener 2000). These treaties envisage a division between a substantial common base, on the one hand, and a set of open partnerships in particular policy sectors, on the other. The common base is defined to include the direct regulation of the single market through a coordinated set of policy measures. Beyond that common base, however, variation is organized sector by sector. The British and Danish opt-outs in respect of the final stages of Economic and Monetary Union or the provisions of enhanced cooperation between interested EU countries in Justice and Home Affairs and Common Security and Foreign Policy are examples of such differentiated integration. Neil Walker has argued that differentiated integration will mean that we may have to envisage "a number of different 'Europes,' their breadth and depth dependent upon the integration arrangements specific to the policy field in question, and embedded in a complex network of relations with one another and with the various Member States" (1998, 356).

The internal fragmentation of nation-states, which manifests itself in regional and ethnonational movements, for example in Spain, Italy, Belgium, or the United Kingdom, adds to this complexity. These subnational entities have been demanding unmediated representation in Europe—indeed, they aim to create a Europe of the regions rather than a Europe of nation-states (never mind a supranational Europe) and hence to construct yet another Europe, politically organized within the institutional structure of the EU in the Committee of the Regions.

Yet, politically organized Europe transcends the European Union. Further political layers to the European polity can easily be added by taking account of the accelerating processes of political regionalization. There are at least two aspects to this process. First, the European Union as a hegemonic power has set political and economic rules that have become accepted by states that are, as yet, outside its political-legal structure. The states of central and eastern Europe (and possibly beyond, as in the case of Turkey) have become part of a European regional system. The systemic interactions between the hegemonic core and the regional peripheries—as they are played out, for example, in the politics of EU enlargement—are yet another element in the European political architecture. Second, in addition to the EU Committee of the Regions, the political mobilization of European regions and municipalities has become organized through the Council of Europe's Congress of Local and Regional Authorities (CLRAE), which was created in 1994 to allow subnational entities to participate in the Council's decision making. There is also the Council of European Municipalities and Regions (CEMR) and the Assembly of European Regions (AER), both of them aiming to strengthen the influence and representation of regions, counties, and local authorities within the supranational European institutions. The Association for European Border Regions (AEBR), finally, aims to politically organize events regarding cross-border problems, help to solve cross-border problems, and prepare and implement common campaigns (Jönsson, Tägil, and Törnqvist 2000, 144–50).

Finally, the European political space is also organized and structured by a plethora of international organizations and regimes. Their variable compositions make for yet another (politically constituted) Europe. While the Council of Europe, the Western European Union (WEU), and the European Free Trade Association (EFTA) have only European members, the Organization for Security and Co-operation in Europe (OSCE) and the North Atlantic Treaty Organization (NATO) also have non-European members. Yet another Europe is constituted, for example, by the signatories of the European Convention on Human Rights.

Clearly, a new political architecture and governance structure in Europe is emerging—both within and outside of the European Union. How can these transformations of the European political space best be conceptualized? I wish to suggest that some analytical advances can be made by taking a historical-comparative perspective on the changes. However, this necessitates not narrowing our sight to the processes that led to the formation of national states in Europe, but focusing on the historically much messier political form of empire.

Empire as a Mode of Governance

More than twenty years ago, Hedley Bull extrapolated from trends then observable in the interstate system and suggested that developments reminded him of the structures and organizational patterns in Christian Europe in the Middle Ages with its twin features of overlapping authority and divided loyalty:

> In Western Christendom in the Middle Ages . . . no ruler or state was sovereign in the sense of being supreme over a given territory and a given segment of the Christian population; each had to share authority with vassals beneath, and with the Pope and (in Germany and Italy) the Holy Roman Emperor above. . . . If modern states were to come to share their authority over their citizens, and their ability to command their loyalties, on the one hand with regional and world authorities, and on the other hand with sub-state or sub-national authorities, to such an extent that the concept of sovereignty ceased to be applicable, then a neo-mediaeval form of universal political order might be said to have emerged. (1977, 254–55)

Hendrik Spruyt, too, emphasized the overlapping allegiances and multiple and competing loyalties in medieval Europe:

> As with all forms of political organization, feudal authorities occupied a geographical space. But such authority over territorial areas was neither exclusive nor discrete. Complex networks of rival jurisdictions overlaid territorial space. Church, lords, kings, emperor, and towns often exercised simultaneous claims to jurisdiction. Occupants of a particular territorial space were subject to a multiplicity of higher authorities. . . . There was no one actor with a monopoly over the means of coercive force. (1994, 12)

Lacking the space to fully substantiate my argument here, let me state that I find the reference to the medieval polity as a governance structure analogous to contemporary formations inadequate. First, for this reference to make any historical sense (along the lines that the modern territorial, sovereign state is postmedieval), we would have to accept the premise of the decline-of-the-state-in-the-age-of-globalization argument. To this argument I do not subscribe (see Axtmann and

Grant 2000, 49–53, for a brief critical discussion of this argument). The state will not wither away but will remain an essential component in the emerging complex governance structure. Even if the state were to lose its monopoly of internal and external violence, it would still be a much more coercive organization than any of the political forms in the Middle Ages. Second, Latin Christendom as an ideology and the papacy as a universal organization gave the Middle Ages a degree of cultural unity that the Reformation, secularization, and nationalism have fractured and replaced with cultural heterogeneity.

Third, capitalism is here to stay—at least for the foreseeable future, making comparisons with the Middle Ages rather fanciful. More specifically, as Hirst and Thompson point out, in the Middle Ages, "[t]he degree of division of labour and economic interdependence was relatively low, whereas today communities depend for their very existence on the meshing and coordination of distinct and often remote activities" (1999, 269). Markets, they argue, can only provide such interconnection and coordination "if they are appropriately governed and if the rights and expectations of distant participants are secured and sustained."

> Hence governing powers cannot simply proliferate and compete. The different levels and functions of governance need to be tied together in a division of control that sustains the division of labour. . . . The governing powers (international, national and regional) need to be "satured" together into a relatively integrated system. . . . The nation-state is central to this process of "saturing": the policies and practices of states in distributing power upwards to the international level and downwards to subnational agencies are the ties that will hold the system of governance together. (269–70)

There may be a plurality of authority structures within and between states, but for authority to be effective, so they claim, it "must be structured by an element of design into a relatively coherent architecture of institutions" (270). If we accept these arguments (and I think there are good reasons to do so), then referring back to the medieval past does not provide us with the analytical insights to gain an understanding of contemporary transformations and future structures.

Therefore, when we take a historical perspective, we should not be looking back to the Middle Ages; we should turn to another historical polity to learn something about the structure and dynamics of governance formations in which the state is not hegemonic and where sovereignty has not been monopolized and is undivided. I suggest that, as a historically unique polity, the Holy Roman Empire of the German Nation in the period after the Peace of Westphalia in 1648 until its dissolution in the early nineteenth century offers comparative-analytical insights into the structure and dynamics of contemporary transformations. We may ask, Does the restructuring of European political space in the twenty-first century resemble the organizational logic of the Holy Roman Empire?

Turning our attention to a polity of the nature of the Holy Roman Empire leads us to a fresh assessment of earlier attempts at organizing supranational polities, and thus to forms of political organization that have been historical alternatives to unitary nation-states. H. G. Koenigsberger (1986) used the expression "composite state"; J. H. Elliott (1992) spoke of "composite monarchies." Elliott quotes Koenigsberger's assertion that "most states in the early modern period were composite states, including more than one country under the sovereignty of one ruler" (50). Elliott makes two sets of distinctions regarding "composite monarchies." First, following Koenigsberger, he says that composite states may either be separated from each other by other states or by sea, like the Spanish Habsburg monarchy, the Hohenzollern monarchy of Brandenburg-Prussia, and England and Ireland; or they can be contiguous, like England and Wales, England and Scotland, Piedmont and Savoy, and Poland and Lithuania. The second distinction goes back to the seventeenth-century Spanish jurist Juan de Solórzano Pereira, who spoke of two ways in which newly acquired territory might be united to a king's dominion. One is *accessory union,* "whereby a kingdom or province, on union with another, was regarded juridically as part and parcel of it, with its inhabitants possessing the same rights and subject to the same laws" (Elliott 1992, 52). The incorporation of the Spanish Indies into the crown of Castile, or the incorporation of Wales into England (Acts of Union of 1536 and 1543) may serve as examples.

But there is also a second form of union, known as *aeque principaliter.* In this union, the constituent kingdoms continue to be treated as distinct entities, and these kingdoms, so Solórzano stated, "must be ruled and governed as if the king who holds them all together were king only of each one of them" (quoted in Elliott 1992, 53). Elliott provides as an example most of the kingdoms and provinces of the Spanish monarchy—Aragon, Valencia, the principality of Catalonia, the kingdoms of Sicily and Naples, and the different provinces of the Netherlands.

At issue is an understanding of the historical experience with complex governance structures within composite supranational polities. These composite polities come in different shapes and sizes. One of them is empire (see Muldoon 1999 for a general discussion on which the following observations draw). The word *empire* has had, of course, a variety of meanings. In the ancient Roman world, one meaning of the term *imperium* was power, specifically the legal power to enforce the law, and thus it did not carry any territorial connotation. By the time of Augustus, however, *imperium* was coming to be understood in two ways. First, the *imperium populi Romani* was the power Romans exercised over other peoples. Empire was viewed as a political entity, comprising the *populus Romanus* and its *socii et amici*. But, second, it was also used in a territorial sense to describe newly acquired lands, lands that had become subjected to the *imperium* of the Roman people.

In the Middle Ages, *imperium* retained both the territorial and nonterritorial connotations. Since the imperial coronation of Charlemagne in 800, the meaning of empire and the political conflict over empire developed around three relationships. First, the relationship between (the initially Frankish) emperor and the papacy; second, that between the Western Roman empire and East Rome (Byzantium); and third, that between universal emperor and individual rulers and kingdoms. The revival of Roman law in the eleventh and twelfth centuries reintroduced the notion of the emperor as *dominus mundi,* thus allowing for a justification of Roman imperial power without reference to the papacy or to Christianity. The twelfth century (1157) also saw a redefinition of the nature of the empire as the Sacrum Romanum Imperium. In the fifteenth century (1474), the empire was renamed as the Sacrum Romanum Imperium Nationis Teutonicae, thus limiting it to the territories ruled by the German monarch.

The term *imperium* was also used more widely. It could be used to refer to rule over kingdoms, and various medieval rulers, the kings in England and Spain, for example, were occasionally referred to as emperors because they had conquered neighboring lands and brought them under their *imperium*. Finally, *imperium* also came to acquire a meaning approximating the modern term *sovereignty.* This is the meaning in the Act in Restraint of Appeals during the rule of Henry VIII in 1533, which states, "This Realm of England is an Empire," thus not only claiming independence from the pope but also reasserting the crown's ambition to extend its territorial authority at the expense of its neighbors. *This* concept of *imperium* goes back to Pope Innocent III's decretal *Per Venerabilem* in the twelfth century and was the anglicized version of the sentence "Rex in suo regno est imperator in regno suo."

It is now often suggested that, from the thirteenth century onward, the history of political thought and political institutions can best be understood as a conflict between two forms of government, a comparatively new one, the emerging modern state, and an old form in the process of transformation, that is, imperial government. Yet, as James Muldoon has rightly argued, such a view is much mistaken:

> To assume that in political terms medieval Europe was some kind of universalist political society while early modern Europe was a collection of sovereign states, each controlling a specific, limited, territory misses the point that it was in the early modern era that European explorers, traders, missionaries, and others were gaining control of lands in the Americas, Asia, and elsewhere, creating, among other problems, a significant problem of governance. How are these newly acquired lands to be ruled? What would be the relation of the European monarchs to their subjects who chose to settle in the newly discovered lands? Would the indigenous inhabitants of the new worlds be treated as the equals of the ruler's European subjects? (1999, 6–7)

At the same time when the European monarchs were trying to transform their medieval dynastic possessions into states domestically, they were acquiring overseas territories that they treated as if they were dynastic possessions.

This is the background of Solórzano's discussion. Let us take Philip II of Spain (1556–98) as an example. Within Spain, his power was limited by local traditions and institutions. He was, after all, king in both Castile and Aragon, and his powers were different in each kingdom. Each was in fact a small empire. They were the products of previous consolidation that had brought the crown of Leon and Granada under the Castilian crown and those of Valencia and Majorca under the Aragonese crown. Furthermore, just as the Castilian crown possessed an empire in the Atlantic, so the Aragonese crown possessed one in the Mediterranean: "[I]n a sense, Philip was the emperor of two empires within Spain itself" (Muldoon 1999, 119). But then, of course, he also possessed the Spanish Netherlands (which he eventually lost) and had obtained the Portuguese throne in 1580. This last acquisition enabled him to claim exclusive rule over all of the newly discovered lands covered by Pope Alexander VI's bull *Inter caetera*.

The Austrian monarchy, or empire, was ruled by the other line of the House of Habsburg. Well into the twentieth century, Austria was an agglomeration of countries and territories. At the end of the fifteenth century, the administrative reforms of Emperor Maximilian I created two major administrative units, Lower and Upper Austria. Lower Austria included the Lower and Upper Austria of today, although only approximately within their current boundaries. But it also incorporated Styria, Carinthia, and Carniola, which together formed Inner Austria. Upper Austria then comprised Tirol, Vorarlberg, the Windisch Mark (roughly, the Slovene territories between Styria and Carniola), Gorizia, Istria, Trieste, and the noncontiguous southwestern German domains, the Vorlande or Vorderoesterreich. This complex of countries and territories (called the ***altösterreichischen Länder***) was augmented in 1526–27 when the lands of the Bohemian crown (Bohemia, Moravia, and Silesia), as well as the crown of Hungary and Croatia, fell to the Habsburgs.

The Austrian monarchy, which was thus formed in the first half of the sixteenth century, was, until its collapse in 1918, a confederation of states *(Staatenverbindung)* (see Kuzmics and Axtmann 2000, 81–133). This meant that, on the one hand, there was a plethora of political constitutions, privileges, immunities, traditions, and customs throughout the monarchy. It also meant that, on the other hand, there existed "[t]echnically no common, but only individual, allegiances of various political domains to the Habsburg sovereignty" (Kann 1974, 3). The political integration of this vast array of territories was the major challenge the Habsburg rulers faced. Would, or should, such an integration mean the centralization of political power in the imperial or royal court and bureaucracy? Or could such integration succeed within the confines of a confederal monarchy? Or could the historical particularisms of the constituent ***Königreiche und Länder*** be acknowledged, or even protected, by the Habsburg rulers within a more federal, and less confederal, union? In the terminology of Solórzano, could, or should, the monarchical union be transformed into an accessory union, or should it be organized as a union ***aeque principaliter?***

■ The Governance Structure of the Holy Roman Empire of the German Nation, 1648–1806

Like the Austrian monarchical union, the institutional order of the Holy Roman Empire of the German Nation was highly complex. Possibly even more so, since, after all, parts of the Austrian monarchical union were also part of the Holy Roman Empire, and, after Charles V, the monarchical rulers in Habsburg Austria were also the emperors of the Holy Roman Empire (with the one exception of Charles VII, who was a member of the Bavarian Wittelsbach house). The Imperial Court in Vienna also functioned as the major meeting place between the nobility from the Austrian hereditary lands, the lands of the Bohemian crown, and those of the Hungarian crown, on the one hand, and the nobility from the Reich, on the other. This was due, first, to the presence of imperial government offices and administration in Vienna (the Imperial Aulic Council and the Imperial Court Chancery, for example). It also arose, second, from the clientelistic politics organized by the emperor, which pulled Austrian and imperial nobility into the court. Finally, this interlinkage between the Austrian and the imperial nobility was also the result of landownership by the imperial nobility in Austria and vice versa as well as intermarriage.

By the late fifteenth century, the core area of the empire covered the territory of modern Germany and Austria, as well as Luxembourg and Belgium; and both the Netherlands and Switzerland were still formally—and until the Westphalian Peace Treaties in 1648—within the empire. The area of the modern Czech and Slovak Republics, then known as the kingdom of Bohemia, with its dependencies of Moravia, Lusatia, and Silesia (the latter now part of Poland), were also within the empire, as were Lorraine, Alsace, and other areas to the west, of which some were lost to France in 1648 but which are now all part of France. The principalities and cities of northern Italy, constituting a region known as Imperial Italy and stretching from Savoy in the west to the frontiers of the Venetian Republic in the east and those of the Papal States to the south, were also part of the empire.

From about the mid–seventeenth century until the collapse of the empire in the early nineteenth century, this territorial complexity manifested itself politically in that approximately 25 major and medium princes, 260 or so minor rulers, counts, and prelates, 50-odd imperial cities, and about 1,500 imperial knights jostled with each other for political power and influence. In general, the emperor would find political (and military) support in the less powerful *(mindermächtige)* estates *(Reichsstände)* against the major and medium princes who wished to curtail his power: for the *mindermächtige* estates, the emperor was the guarantor of their autonomy against the predatory designs of the more powerful territorial princes. At the same time, the small group of seven to nine major princes who had the constitutionally enshrined right to elect the emperor, the electors *(Kur-*

fürsten), often sided with the emperor to fend off the demands for more participation and political influence by the other princes.

Manifestly, the governance structure of this conglomerate of political units was complex, conflictual, and contested. This is not the place to provide a comprehensive picture of the Holy Roman Empire as a political-legal formation sui generis. Yet, in the context of our discussion on the emerging new political architecture in Europe, some discussion of certain key aspects of governance in the Holy Roman Empire after 1648 is warranted (see Wilson 1999; Aretin 1993; Schmidt 1999; Press 1997, 1998; Schindling 1991; Dotzauer 1998 for general surveys and detailed bibliographies).

The Reichstag (Imperial Diet) was one of the empire's key institutions. In its final form the Reichstag had three colleges, or *curia.* The first was the college of electors *(Kurfürstenrat),* consisting of between seven and nine electors. The second *curia* was the college of princes *(Fürstenrat).* It was composed of princes, imperial counts, and prelates. Whereas the princes possessed full individual votes (*Virilstimmen* [in 1792, there were 100 votes]), the counts and prelates shared collective votes *(Kurialstimmen):* the prelates possessed two collective votes and the counts four votes. The third *curia* was the college of imperial cities *(Städtekurie)* with fifty-odd votes.

The competencies of the permanent Imperial Diet *(Immerwährender Reichstag)* after 1663 were comprehensive. It had wide-ranging legislative powers in domestic matters, mainly with regard to internal order, peace, and security as well as the regulation of the economy; but it also, in effect, was the supreme interpreter of imperial law. It possessed the right to approve of taxation. Questions of war and peace, and all related political-military matters, were also part of its remit.

The institutionalization of the Diet had a number of prerequisites. First, as an institution, the Imperial Diet found support both from the less powerful estates and the emperor. For the less powerful estates, the Diet offered the opportunity of participation in political rule. For the emperor, the Diet allowed the forging of political alliances with the *mindermächtige* estates against the hegemonic design of the larger territorial princes. But for (at least) some of the powerful princes, too, the Diet was an attractive option. Above all, the small group of electors could exercise regular and permanent influence on political decision making. After all, decision making in the Diet was organized through majority decisions in each *curia.* Once a decision had been reached by the college of electors and the college of princes, the two colleges would then agree on a position, which they would present to the college of imperial cities, either for further negotiation or for joint agreement. Such an agreement would then need the emperor's approval to become imperial law and hence binding on all members of the *Reich*. This procedure ensured a high degree of political influence for the numerically small *curia* of electors. Second, the constitutional settlement of 1648, and with it the institution of the Imperial Diet, was guaranteed by two foreign powers, France and Sweden. For geopolitical reasons, each power had an entrenched interest in seeing a system sustained that was organized around the institutional and geographical dispersal of political power. Furthermore, three foreign powers were themselves represented in the Imperial Diet. Sweden, Denmark, and, after the Hannoverian succession, England possessed territory in the Holy Roman Empire that made them imperial estates.

Third, for the Diet to function relatively efficiently and effectively, there had to be a political will for compromise and a political culture of tolerance. Given the history of religious conflict that had characterized politics in the Holy Roman Empire since the early sixteenth century, such a willingness for compromise and tolerance would have to manifest itself in dealing with the religious conflict. And indeed, following provisions made in the Westphalian Peace Treaties, the Imperial Diet instituted procedures to manage religious divisions. In those cases where either the Protestant or Catholic members of the Diet asserted that policy decisions would involve religious matters, the Diet could constitute itself in two *corpora,* a Protestant and a Catholic *corpus (itio in partes)*. Decisions could then no longer be taken through majority vote, but would have to be based on an amicable and negotiated search for a compromise ("sola amicabilis compositio divinat") (IPO, Art. V, § 52). This consociational procedural mechanism could very easily have led to the paralysis of the Imperial Diet. Yet, during the 158 years of its existence, the procedure was used only eight times. This would indicate that both religious groupings, the Catholic and the Protestant (Lutheran and Reformed) *corpora,* perceived the Diet as an institution that was not systemically or of necessity detrimental to their interests.

It was in the Imperial Diet that the particularistic interests of the individual *Reichsstände* could clash with each other and where, between conflict and consensus, common policies for the empire as a whole had to be thrashed out. It was there where communication flows converged and information was exchanged and where a political public sphere could form. As a result, the Imperial Diet would gradually frame imperial, or translocal and national, politics in Germany.

The highest judicative power in the empire was exercised by two imperial courts, the Reichshofrat (Imperial Aulic Council) and the Reichskammergericht (Imperial Cameral Court). Located at the Imperial Court in Vienna, the Imperial Aulic Council was both a judicial body and an advisory council to the emperor. The Imperial Cameral Court, however, was located far away from Vienna in Speyer and later in Wetzlar, and the emperor's influence on this body was weaker than that of the Imperial Diet and the Imperial Estates; after all, whereas the Aulic Council was financed by the emperor himself, the Cameral Court depended on the financial support of the estates.

Like all imperial institutions, the two courts faced attempts by the powerful territorial princes to curtail their powers and competencies, whereas the ***mindermächtige Stände*** wanted to be able to avail themselves of judicial redress in these courts. Both courts dealt with essentially the same matters: conflicts between territorial rulers; conflicts within territories, especially those between rulers and their estates; and conflicts arising out of social and religious conflicts and tensions. Of considerable importance for the subjects of rulers was their right to appeal to either of the supreme courts against judgments issued by the rulers' courts. While such appeals could only be lodged in a clearly limited number of cases—and both the electoral princes and many other princes had obtained the ***privilegium de non appellando,*** barring subjects from appeals to the imperial courts—the courts contributed to a judification of the empire and the awareness among the Imperial Estates and subjects alike that the ***Reich*** was a constitutional-legal order encompassing all of its members.

One main problem was the difficulty of enforcing the verdicts of either court. But the Imperial Diet, too, faced this problem: who would execute its policy decisions? While the Diet was the highest legislative body in the empire and the courts the highest judicial bodies, the Imperial Circles *(Reichskreise)* were the most important administrative and executive bodies. From 1512 until the early nineteenth century, the empire had been divided into ten Imperial Circles. These circles did not necessarily have a contiguous geographical territory; they could contain enclaves from other circles on their own territory, which means also that many circles had exclaves in other circles. One of their main tasks was to execute the decisions taken by the Diet, which, however, were frequently only formulated in general terms and were to be turned into specific actions by the circle authorities on the basis of due consideration of local circumstances. Their functions were comprehensive. Besides the maintenance of public peace, the selection of judges for the Imperial Cameral Court, the execution of the judgments made by the Imperial Courts, the collection of taxes, the coordination of the provision of territorial military contingents for national defense (i.e., the imperial army) and internal peacekeeping, and the supervision of coinage regulations, after 1648, the circles were increasingly involved in all police matters, which ranged from public health to public welfare and public infrastructure (e.g., road building).

In each of the Reichskreise, one to three important princes acted as the executive heads *(Kreisausschreibender Fürst)*. Counterbalancing the power of the executive princes, *Kreis* assemblies were established, which, in their structure, took the Imperial Diet as their example. However, the votes in the *Kreis* assemblies had an equal value. From the point of view of our discussion of governance structures, three important points should be made. First, neither the emperor nor the Reichstag had any influence on the convocation of the *Kreis* assembly nor on its deliberations. Second, not all of the Imperial Circles were very active. Wilson (1999, 58) is correct to state that, as a general rule, "[t]hose Kreise which contained a greater proportion of weaker territories tended to be the most active, because the Kreis structure to an extent substituted for the administrative, political, military and fiscal infrastructures that these petty potentates were unable to develop themselves." But we should also recognize that the more powerful territorial princes situated within a circle in many cases did not wish to see it built up as an alternative political structure to their own political administration. They opposed this institution just as they opposed other elements of the imperial constitution and, unintentionally, provided the emperor with a chance for more influence within the circles. Third, in order to discharge and expedite their tasks efficiently, *Kreis* authorities met to coordinate their activities. Such cooperation was particularly well institutionalized in the area of coinage regulation *(Münzprobationstage)*. But the Westphalian Peace Treaties had also reinforced the right of the *Kreise* to enter into associations and alliances as long as they were not directed against the peace or the *Reich*. As a result, a considerable number of associations between *Kreise* were formed, mainly in the realm of military defense but also in the area of public order and the maintenance of good police.

The corporation of the Imperial Knights shall serve as a final example for the multilayered governance structure of the Holy Roman Empire. The Imperial

Knights had managed to remain outside both the Reichstag and the *Kreis* structure; instead, they had submitted themselves directly to the emperor in a clientelistic relationship. Since 1577, the Imperial Knights were organized in a three-layered corporative structure. There were the three Swabian, Franconian, and Rhenish circles *(Ritterkreise)*, each one an autonomous knightly corporation with its own specific constitution and specific privileges and liberties granted by the emperor. Each *Ritterkreis* had its own executive officers and a deliberative *Kreis* assembly. The second level was constituted by the cantons into which each circle was subdivided, each of them also organized by an executive and a cantonal assembly. At the third level, the Imperial Knights were organized translocally through a general executive *(Generaldirektorium)* and a general assembly *(Generalkorrespondenztage)* that coordinated the activities of the three circles. The tasks of the *Ritterkreise* and the *Ritterkantone* were very similar to those of the Reichskreise, in particular with regard to taxation (which remained a right of the cantons), public order, and public welfare (police). Nested as they were geographically within the Imperial Circles, the relationship between the corporative Imperial Knights and the Imperial Circles was always strained and in need of negotiated compromise. It was through this form of cooperation that a further intensification of governance could be achieved.

Peter Wilson (1999, 8–16) has analyzed well the dynamics and structural possibilities of the polity of the Holy Roman Empire. (1) The *Reich* could have been transformed into a state, institutionalizing greater direct imperial control over the empire by depriving the princes and lesser rulers of all or part of their autonomy through some form of imperial absolutism. (2) The collective-corporate elements of imperial institutions like the Imperial Diet or the Imperial Cameral Court could have been strengthened, thus allowing for a continued evolution of the traditional hierarchical structure that gave the empire its peculiar character of a multilayered polity. This was characterized by the emperor's overall authority, while disallowing his direct control as a result of the constitutionally entrenched right of the Imperial Estates of political representation and co-decision-making. (3) The growth of federalism was the third political tendency alongside the monarchical principle and the traditional hierarchy. This federalism (could have) developed in three forms. First, there was *princely-territorial federalism* as a result of the emergence of distinct territories and state formation. The princes pursued absolutist centralization within their own territories, while resisting it on the part of the emperor: "[T]o forestall any moves towards more direct imperial rule, the territorial rulers tended to champion their right to representation at national [i.e. *Reich,* RA] level through the estates principle" (Wilson 1999, 12). As a result of this strategy, the intermediate levels between emperor and territories, such as the Reichstag or the regional subdivision of the empire into Reichskreise were strengthened and consolidated.

This was one reason for a centripetal federalism. However, the fact that princely power was legitimated from above by imperial law and its immediate relationship to the emperor, as well as the fact that few of the German territories possessed the potential to survive as viable, independent states outside the imperial structure, further enhanced the federal orientation. The result was a structural dynamics in favor of the formation of a loose federation of autonomous, consolidated states where the tendency toward dissolution or exit from the federal structure could be contained. In addition to these federal trends, there was, secondly, ***cross-regional political cooperation among the nobility,*** such as those of the Imperial Knights, and aristocratic confederations between estates of different territories. Finally, there were also ***confederations and alliances between cities and peasant communities,*** which added a further layer of federal organization.

Wilson has succinctly summed up these trends and tendencies:

> Monarchical centralization remained present, and though largely displaced to the level of the princely territories, continued to offer an alternative in the form of "imperial absolutism" into the seventeenth century. Thereafter, the need to coordinate action between the territories could still strengthen the emperor's position, since he remained the obvious focal point for common policy. By reinforcing the cohesion of the larger territories, princely centralization also reinforced the federal tendency within the Empire and so undermined the traditional hierarchical structure in the long run. Alliance between princes cut across constitutional structures and pushed the Empire towards a looser association of equal states. . . . Yet . . . federalism also contained strains inhibiting territorialization and which could strengthen the imperial constitution. For instance, the Kreis Association movement contained elements that were federal, and by preserving the weaker rulers as active players in imperial politics, it prolonged the life of the constitution. Popular and aristocratic forms of federalism also slowed the growth of princely power and had constituted alternative forms of government at important phases of the Empire's history. (1999, 70–71)

In the Westphalian Peace Treaties of 1648, all electors, princes, and estates of the Holy Roman Empire had their old rights, prerogatives, liberties, and privileges confirmed together with their *superioritas territorialis* in all matters spiritual and secular ("libero iuris territorialis tam in ecclesiasticis quam politicis exercitio") (IPO, Art. VIII, §1). This *ius territorialis* included the right to levy taxes, to legislate for their subjects, and to raise an army. Article VIII, §2, also confirmed their right to enter into alliances among themselves and with foreign powers to ensure their own preservation and security, as long as such alliances would not be directed against the emperor, the empire, or the peace treaty. As a consequence of this provision, the Imperial Estates become subjects in international law. While these two provisions would allow the larger and more powerful territorial princes to claim (and achieve) sovereignty in the course of the eighteenth century, *ius territorialis* was not the same as *souveraineté* à la Jean Bodin or Thomas Hobbes. Only if, and as long as, territorial laws did not contradict imperial legislation, did the rule of nonintervention and noninterference apply. To that extent *ius territorialis* did not constitute a highest, ultimate authority but retained the principle of the subjection of all members of the Holy Roman Empire to the emperor and the institutions of the *Reich*. The two supreme courts, the Imperial Aulic Council and the Imperial Cameral Court, did not only guarantee the subjects of the territorial rulers, that is, of the Imperial Estates, protection against arbitrary treatment and perversion of justice, but they also supervised and scrutinized territorial legislation. Through their norm-setting rulings, the Supreme Imperial Courts reinforced the limits to *ius territorialis* and erected a barrier against the appropriation of full sovereignty by the estates. Why such a system of shared *superioritas* between "Kaiser und Reich" and individual estates unraveled in the second half of the eighteenth century must remain outside the concern of this study.

■ Conclusion

I argued that the Holy Roman Empire offers rich material for an analysis of the conflictual relationships between centralization, federalism, and devolved government and the institutionalization of (shared) sovereignty. For this reason, we may well gain a better understanding of the contemporary reconfiguration of political forces in Europe by taking another look at a polity that, allegedly, was left behind as outmoded by the triumphant sovereign territorial state. These suggestive observations are not limited to the European Union as a political community but are meant to illuminate processes of reconfiguration in the European political space—which does witness both a restructuring of political authority within the EU (vide the European Convention, which is intended to lay the foundation for a European constitution) and an endeavor to rearticulate the political organizational structures and relations of the EU with its European political environment.

Nevertheless, it is noticeable that in the academic debate on the future of the European Union, and its potential for further democratization in the period of its enlargement, reference has been made—even if it has not been fully acknowledged—to the experience of the Holy Roman Empire. Philippe C. Schmitter (2000) has recently suggested grouping the countries of an enlarged European Union into three "colleges" according to size in population, and he linked this institutional innovation to the idea of "concurrent majorities." As he explains:

> The basic principle is simple: No measure would pass the Council of Ministers unless it obtained a majority in each of the three *colegii*—a simple majority for ordinary directives and perhaps a qualified majority for such matters as amendments to the rules or admission of new members. This should be sufficient to reassure the smaller countries that they will not be outvoted *as a category* by the larger ones, although each country would still face the imperative of convincing a sufficient number of the fellow members of its *colegio* to reject a measure that it regarded as a specific threat to its national interests. (84–85)

As should by now be clear, this idea is close in spirit to the institutional arrangements in the Imperial Diet.

Yet, while a dispassionate look at the constitution of the Holy Roman Empire may yield an interesting analytical perspective, we must not forget that concrete political history and passions propel the reconfiguration of the European political space. In May 2000, Jean-Pierre Chevènement, then the French interior minister, said that "there is a tendency in Germany to imagine a federal structure for Europe which fits in with its own model. Deep down, it is still dreaming of the Germanic Holy Roman Empire. It hasn't cured itself of its past derailment into Nazism."[1] Arguably, Chevènement was more concerned with reminding his listeners of the geopolitical rivalry between France and the Holy Roman Empire than with giving due regard to the constitutional complexities of the Holy Roman Empire. Nonetheless, in the light of such remarks, it bears repeating that I have suggested in this chapter that the Holy Roman Empire may serve as an analytical, but not as a political, model for understanding the reconfiguration of the European political space.

NOTE

1. *The Guardian,* 23 May 2000, 18.

SIX

Rescaling State Space in Western Europe

Urban Governance and the Rise of Glocalizing Competition State Regimes (GCSRs)

NEIL BRENNER

State theorists and political geographers have long emphasized the territorial properties of political power in the modern world. Within the Westphalian geopolitical order, states are said to be composed of self-enclosed, contiguous, and mutually exclusive territorial spaces that separate an *inside* (the realm of political order, citizenship, and/or democracy) from an *outside* (a realm of interstate violence and anarchy) (Walker 1993). For the most part, however, even while being acknowledged as an underlying feature of modern geopolitical organization, territoriality has been treated within mainstream approaches to political studies as a relatively fixed and inconsequential property of statehood. Just as a fish is unlikely to discover water, it has been argued, most postwar social scientists viewed national state territories as pregiven natural environments for social life (Taylor 1994, 157). A "territorial trap," to use Agnew's memorable phrase (1994), has long underpinned mainstream approaches to political studies, insofar as they have conceived state territoriality as a static background structure for regulatory processes and political relations rather than as one of their constitutive dimensions.

During the last decade, however, in close conjunction with ongoing debates on globalization and the crystallization of a "post-Westphalian" world order, these entrenched methodological assumptions have been thoroughly unsettled. As many scholars have noted, the global political-economic transformations of the post-1970s period have dramatically reconfigured the Westphalian formation of state territoriality, (1) by decentering the national scale of state regulatory activity and (2) by undermining the internal coherence of national economies and national civil societies as targets for state policies (Agnew and Corbridge 1995). Under these conditions, scholars throughout the social sciences have become attuned to the dynamically changing geographies of statehood within contemporary global capitalism. Although our understanding of the new state spaces that are currently emerging remains rudimentary, recent research on geopolitical and geoeconomic restructuring has provided some fruitful methodological starting points through which the geographies of state power of the early twenty-first century may be explored more systematically (for an overview of these emergent lines of research, see Brenner et al. 2003).

Against the background of these debates, this chapter develops an interpretation of state spatial restructuring in contemporary western Europe, where the consolidation of the European Union has underpinned dramatic transformations of political space. While supranational state restructuring processes — such as the institutionalization of Europe-wide regulatory arrangements — are of profound significance, the following discussion focuses primarily upon the transformation of state spatiality on national and subnational scales. The phenomenon of state ***downscaling*** — the devolution or decentralization of regulatory tasks to subnational administrative tiers, coupled with a restructuring of subnational institutional spaces through various national, regional, and local policy initiatives — is arguably as fundamental to the contemporary remaking of European political space as the processes of state ***upscaling*** that have been examined by radical critics of neoliberal forms of globalization. I shall propose, in particular, that the remaking of ***urban governance*** represents a key mechanism through which processes of state rescaling are unfolding throughout the EU.

For present purposes, the concept of urban governance refers to the broad constellation of social, political, and economic forces that mold the process of urban development within modern capitalism. Urban governance occurs on a range of geographical scales insofar as the process of capitalist urbanization has expanded to encompass not only individual cities and towns but also large-scale urban regions, cross-border metropolitan agglomerations, national city-systems, and supranational urban hierarchies. State institutions are hardly the only important influence on urban governance. Capitalist firms, property owners, business organizations, commercial interests, utilities suppliers, trade unions, and diverse social movements may likewise play key roles in shaping urban fortunes under capitalism (Logan and Molotch 1987). Nonetheless, it can be argued that the state's impact upon urban development processes has expanded considerably since the consolidation of organized capitalism during the early twentieth century.

It was during this era, and particularly after the Second World War, that western European states began to invest extensively in the construction of large-scale infrastructures for capital circulation and social reproduction; to engage in increasingly sophisticated, long-term forms of urban, regional, and spatial planning; and to intervene directly in the regulation of capital's uneven geographical development (Lefebvre 1978). Once the state began to operate as the "overall manager of the production and reproduction of social infrastructures" (Harvey 1982, 404) within each national territory, the politics of national economic development were interlinked ever more closely with a variety of direct and indirect state strategies to influence the form, pace, and geographies of urban growth. In this sense, urban governance must be viewed as the key regulatory arena in which states attempt continually to impose a "structured coherence" (Harvey 1989c) upon the process of capitalist urbanization and thus to regulate the uneven development of capital within their territories.

To be sure, urban governance represents only one among many strategic institutional arenas and targets of state spatial restructuring in the EU—monetary regulation, social policy, labor policy, gender relations, and citizenship/migration are likewise highly important, deeply contested political arenas in which a "new *Gestalt* of scale" is being forged in the context of European integration (Swyngedouw 2000). It can be argued, nonetheless, that urban governance provides an illuminating analytical window through which to explore contemporary rescalings of state spatiality and their ramifications for the political-economic geographies of postfordist western Europe. In particular, this chapter argues that political strategies to reconfigure the institutional infrastructure of urban governance have been among the key generative forces underlying the contemporary downscaling of state space throughout the EU.

In order to develop this thesis, I shall argue that *four* successive configurations of urban governance have crystallized in western Europe since the era of high Fordism in the late 1950s and that each of these configurations has been closely intertwined with historically specific strategies of state spatial regulation. I shall suggest, moreover, that each successive configuration of urban governance has been associated with a historically specific, state-led "scale-making project" (Tsing 2000). Such scale-making projects have entailed not only strategies to establish particular scales of state power in which the process of capitalist urbanization may be *regulated,* but also, just as crucially, strategies to *position* cities within a broader, supralocal geography of state institutions and capital flows. To the extent that state institutions may ever successfully establish even a provisional structured coherence for capitalist urbanization, it is through the meshing together of these *city-regulating* and *city-positioning* aspects of urban governance within their territories. As we shall see, the consolidation of qualitatively new scale-making projects in the realm of urban governance during the last two decades has underpinned a fundamental reworking of state spatiality throughout western Europe. From this perspective, the entrepreneurial, development-oriented approaches to urban governance that have proliferated during the post-1970s period represent key mechanisms and expressions of a fundamental transformation of state spatiality throughout western Europe.[1] The socially and spatially redistributive Keynesian welfare national states of the Fordist-Keynesian era are being tendentially superseded by glocalizing competition state regimes (GCSRs) oriented toward a geographical reconcentration of productive capacities and economic assets within strategic urban regions and industrial districts.

■ The 1960s: Geographies of Urban Governance at the High Point of Spatial Keynesianism

The economic geography of postwar western Europe was composed of a functional division of space on various geographical scales (Lipietz 1994). Spatial divisions of labor emerged within each national territory in the form of hierarchical relationships between large-scale metropolitan regions, in which the lead firms within the major, propulsive Fordist industries were clustered, and smaller cities, towns, and peripheral zones, in which branch plants, input and service providers, and other subordinate economic functions were located. In the western European context, the geographical heartlands of the Fordist accumulation regime stretched from the Industrial Triangle of northern Italy through the German Ruhr district to northern France and the English Midlands; but each of these regional production complexes was in turn embedded within a nationally specific system of production. Throughout the postwar period, these and many other major European urban regions and their surrounding industrial satellites were characterized by consistent demographic growth and industrial expansion. As the Fordist accumulation regime reached maturity, a major decentralization of capital investment unfolded as large firms began to relocate branch plants from core regions into peripheral spaces (Dunford and Perrons 1994; Rodríguez-Pose 1998). Under these conditions, urban governance acquired a key role in a variety of *nationalizing* scale-making projects in which western European states attempted to construct centralized bureaucratic hierarchies, to establish nationally standardized frameworks for capitalist production and collective consumption, to underwrite urban and regional growth, and to alleviate uneven spatial development throughout their national territories.

First, in order to standardize the provision of welfare services and to coordinate national economic policies, national states centralized the instruments for regulating urban development, thereby transforming local states into mere "transmission belts" for centrally determined policy regimes (Mayer 1992). Within this managerial framework of urban governance, the state's overarching function on the urban scale was the reproduction of the labor force through public investments in housing, transportation, social services, and other public goods, all of which were intended to replicate certain minimum standards of social welfare across the national territory (Harvey 1989b; Castells 1977). In this manner, local states were instrumentalized in order "to carry out a national strategy based on a commitment to regional balance and even growth" (Goodwin and Painter 1996, 646). Insofar as the national economy was viewed as the primary terrain for state action, local and regional economies were treated as mere subunits of relatively autocentric national economic spaces dominated by large-scale corporations. These centrally financed local welfare policies also provided important elements of the social wage, and thus they contributed significantly to the generalization of the mass consumption practices upon which Fordist growth was contingent (641).

As theorists of the "dual state" subsequently recognized, the pervasive localization of the state's collective consumption functions during the postwar period was a key institutional feature within a broader scalar division of regulation in which production-oriented state policies were organized on a national scale (Saunders 1979). Accordingly, throughout this period, state strategies to promote economic development, including urban economic development, were mobilized primarily on a national scale rather than through regional or local initiatives. In this context, a range of national social and economic policy initiatives—including demand-management policies, the nationalized ownership of key industries (coal, shipbuilding, power, aerospace), the expansion of public-sector employment, military spending, and major expenditures on housing, transportation, and public utilities—served directly or indirectly to underwrite the growth of urban and regional economies (Martin and Sunley 1997, 280).

Second, even though major cities and metropolitan regions received the bulk of large-scale public infrastructure investments and welfare services during the Fordist-Keynesian epoch, such city-centric national state initiatives were counterbalanced through a variety of state expenditures, loan programs, and compensatory regional aid policies designed to spread growth into underdeveloped regions and rural peripheries across the national territory. From the Italian Mezzogiorno and Spanish Andalusia to western and southern France, the agricultural peripheries and border zones of West Germany, the Limburg coal-mining district of northern Belgium, the Dutch northeastern peripheries, the northwestern regions and islands of Denmark, the Scandinavian north, western Ireland, and the declining industrial zones of the English North, south Wales, parts of Scotland, and much of Northern Ireland, each European country had its so-called problem areas or lagging regions, generally composed of economic zones that had been marginalized during previous rounds of industrial development or that were locked into increasingly obsolete technological-industrial infrastructures (Clout 1981; OECD 1976).

Consequently, throughout the postwar period until the late 1970s, a range of regional policies were introduced across western Europe that explicitly targeted such peripheralized spaces. Generally justified in the name of priorities such as "balanced national development" and "spatial equalization," these redistributive regional policies entailed the introduction of various forms of financial aid, locational incentives, and transfer payments to promote industrial growth and economic regeneration outside the dominant city cores; and they often channeled major public infrastructural investments into such locations (Clout 1981; OECD 1976). Such interregional resource transfers had a significant impact upon the geographies of uneven development during the postwar period, contributing to an unprecedented convergence of per capita disposable income within most western European states (Dunford and Perrons 1994). This nationally oriented project of industrial decentralization, urban deconcentration, and spatial equalization was one of the foundations of spatial Keynesianism, the system of state territorial regulation that prevailed throughout the Fordist-Keynesian period (Martin and Sunley 1997).

Third, it is worth emphasizing that, within this nationalized system of urban governance, metropolitan political institutions acquired an important mediating role between managerial local states and centrally organized, redistributive forms of spatial planning. Between the mid-1960s and the early 1970s, diverse types of consolidated metropolitan institutions were established in many major western European city-regions (Sharpe 1995). These metropolitan administrative bodies were widely viewed as mechanisms for rationalizing welfare service provision and for reducing administrative inefficiencies within expanding urban agglomerations. In this sense, metropolitan institutions served as an important coordinating administrative tier within the centralized hierarchies of intergovernmental relations that prevailed within the Keynesian welfare state apparatus. As suburbanization and industrial decentralization intensified, metropolitan political institutions were also increasingly justified as a means to establish a closer spatial correspondence between governmental jurisdictions and functional territories (Lefèvre 1998). By the early 1970s, metropolitan authorities had acquired important roles in guiding industrial expansion, infrastructural investment, and population settlement beyond traditional city cores into suburban fringes, primarily through the deployment of comprehensive land-use plans and other mechanisms to influence intrametropolitan locational patterns. In this sense, metropolitan institutions appear to have significantly influenced the geographies of urbanization during the period of high Fordism.

In sum, urban governance was an essential pillar within the nationalized system of spatial Keynesianism that prevailed throughout much of western Europe from the late 1950s until the early 1970s. Spatial Keynesianism is best understood as a broad constellation of national state strategies designed to promote capitalist industrial growth by alleviating or overcoming uneven geographical development within each national economy (Martin and Sunley 1997). Thus conceived, spatial Keynesianism represented a state-led scale-making project in two senses: first, it entailed the establishment of a complex system of subnational institutions for the territorial regulation of urban development; and second, it entailed the embedding of local and regional economies within a hierarchically configured, nationally focused political and economic geography. Throughout the postwar period, local governments were subsumed within nationalized institutional matrices characterized by centralized control over local social and economic policies, technocratic frameworks of metropolitan governance, extensive interregional resource transfers, and redistributive forms of national spatial planning. Taken together, such policies and institutions attempted to promote a structured coherence for capitalist growth (1) by transforming cities and regions into the localized building blocks for national economic development and (2) by spreading urbanization as evenly as possible across the national territory. By the early 1970s, however, it had become apparent that the fantasy of transcending uneven spatial development through the promotion of balanced urbanization within a relatively closed national economy was as short-lived as the Fordist accumulation regime upon which it was grounded.

■ The 1970s: Crisis Management and the New Politics of "Endogenous" Growth

A new configuration of urban governance and state spatial regulation began to crystallize as of the early 1970s (Lipietz 1994). A number of geoeconomic shifts occurred during this era that decentered the predominant role of the national scale as a locus of political coordination and led to the transfer of new regulatory responsibilities and burdens both upward to supranational institutional forms such as the EU and downward to the regional and local levels. The uneven, incremental, and deeply contested character of this "relativization of scales" (Jessop 2000) is illustrated quite clearly in the realm of urban governance, where the transformation of state spatiality was mediated through a range of regulatory responses, crisis-management strategies, and political experiments.

In a seminal discussion, Lipietz (1994, 35) has underscored the ways in which processes of capitalist restructuring are articulated in the form of struggles between "defenders of the 'old space'" (which he refers to as the "conservative bloc") and proponents of a "new space" or a "new model of development" (which he refers to as "the modernist bloc"). For Lipietz, the production of new spaces occurs through the conflictual interaction of conservative/preservationist and modernizing or restructuring-oriented political forces on diverse scales, generally leading to a new territorial formation that eclectically combines elements of the old geographical order with aspects of the "projected spaces" sought by the advocates of (neoliberal or progressive) modernization. This conceptualization of state spatial restructuring as a conflictual interaction of inherited institutional landscapes and emergent regulatory strategies illuminates important elements of urban governance restructuring during the 1970s. Throughout this decade, intense conflicts between preservationist and restructuring-oriented political blocs unfolded on a range of spatial scales, generating uneven impacts upon the framework of urban governance that had been established during the preceding decades.

On the one hand, on a national scale, conservative/preservationist blocs mobilized diverse strategies of crisis management in order to defend the decaying institutional infrastructures of the Fordist-Keynesian order. From the first oil shock of 1973 until around 1979, traditional recipes of national demand-management prevailed throughout the OECD zone as central governments tried desperately to recreate the conditions for Fordist forms of economic growth. However, as Jessop (1989, 269) remarks of the British case, such countercyclical tactics ultimately amounted to no more than an "eleventh hour, state-sponsored Fordist modernisation," for they were incapable of solving, simultaneously, the dual problems of escalating inflation and mass unemployment. Meanwhile, as the boom regions of postwar Fordism experienced sustained economic crises, the policy framework of spatial Keynesianism was further differentiated to include deindustrializing, distressed cities and manufacturing centers as key geographical targets for various forms of state assistance and financial aid. In contrast to traditional Keynesian forms of spatial policy, which had focused almost exclusively upon underdeveloped regions and peripheral zones, national *urban* policies were now introduced in several western European states to address the specific socioeconomic problems of large cities, such as mass unemployment, deskilling, capital flight, and infrastructural decay. Major examples of such policies included the West German Urban Development Assistance Act, the French Plan of Action for Employment and Industrial Reorganization, the Dutch Big Cities Bottleneck Program, and the British Inner Urban Areas Act (Fox Przeworski 1986). In this manner, many of the redistributive policy relays associated with spatial Keynesianism were significantly expanded during the 1970s. Crucially, however, even though the spatial targets of regional policies were now differentiated to include urban areas as key recipients of state aid, the state's underlying commitment to the project of spatial equalization on a *national* scale was maintained and even reinforced throughout this decade. Indeed, the 1970s can be viewed as the historical culmination of the various projects of national sociospatial redistribution that had been introduced during the "golden age" of Fordist expansion, albeit under geoeconomic conditions that were seriously undermining the Fordist developmental model.

On the other hand, however, even as these new forms of state support for urban development were extended, a range of national policy initiatives and intergovernmental realignments unsettled the managerial-welfarist framework of urban governance that had prevailed throughout the postwar period. As of the late 1970s, the national scale likewise became an important institutional locus for modernizing, restructuring-oriented political projects that aimed to dismantle many of the policy relays associated with the Keynesian welfare national state. During the post-1970s recession, as national governments were pressured increasingly to rationalize expenditures, national grants to subnational administrative levels, including both regions and localities, were generally reduced. These new forms of fiscal austerity caused local governments throughout western Europe to become more dependent upon locally collected taxes and nontax revenues such as charges and user fees (Mouritzen 1992). In the immediate aftermath of these shifts, many western European local governments attempted to adjust to the new fiscal conditions by delaying capital expenditures, drawing upon liquid assets, and increasing their debts; however, these strategies proved to be no more than short-term stopgap measures. Subsequently, additional local revenues were sought in economic development projects (Fox Przeworski 1986), among other sources. Whereas the new national urban policies introduced during this period enabled many cities to capture supplementary public resources, most local governments were nonetheless confronted with major new budgetary constraints due to the dual impact of national fiscal retrenchment and intensifying local socioeconomic problems. One of the most significant institutional outcomes of the national fiscal squeeze of the 1970s, therefore, was to pressure localities to seek new sources of revenue through a proactive mobilization of local economic development projects and inward investment strategies.

Under these conditions, the managerial-welfarist form of local economic governance that had been consolidated during the postwar period was significantly unsettled as bootstrap strategies to promote economic growth from below, without extensive reliance upon national subsidies, proliferated in major western

European cities and regions (Bullmann 1991). In contrast to their earlier focus on welfarist redistribution, local governments now began to mobilize diverse strategies to rejuvenate local economies, beginning with land-assembly programs and land-use planning schemes and subsequently expanding to diverse firm-based, area-based, sectoral, and job-creation measures (Eisenschitz and Gough 1993). Although this new politics of urban economic development would subsequently be diffused in divergent political forms throughout the western European city-system, during the 1970s it remained most prevalent within manufacturing-based cities and regions in which industrial decline had generated particularly devastating socioeconomic problems (Parkinson 1991). Thus, even as national governments continued to promote economic integration and territorial equalization on a national scale, neocorporatist alliances between state institutions, trade unions, and other local organizations within rustbelt cities and regions from the German Ruhr district to the English Midlands elaborated regionally specific sectoral, technology, and employment policies in order to promote what was popularly labeled "endogenous growth" (Hahne 1985; Stöhr and Taylor 1981). Throughout the 1970s, the goal of these neocorporatist and social democratic alliances was to establish negotiated strategies of industrial restructuring in which economic regeneration would be linked directly to social priorities such as intraregional redistribution, job creation, vocational retraining initiatives, and class compromise.

These initiatives were grounded upon an essentially neo-Fordist political project to rescale the institutional infrastructure of spatial Keynesianism from a national to a regional or local scale. Regional and local economies were now recognized to have their own specific developmental trajectories and structural problems rather than being mere subunits within a unitary national economic space. At the same time, however, the basic Fordist-Keynesian priorities of social redistribution, territorial equalization, and class compromise were maintained, albeit within the more bounded parameters of regional or local economies rather than as a project to be extended throughout the entire national territory. Thus, in contrast to the nationally focused redistributional policies mobilized by the central state during the 1970s, the new politics of endogenous growth during this period were oriented toward place-specific regulatory problems, socioeconomic dilemmas, and political conflicts. Many of these neocorporatist local economic initiatives would continue during the 1980s, though in a radically transformed geopolitical and geoeconomic environment. The salient point here is that such strategies of endogenous growth first emerged during a period in which neo-Keynesian priorities continued to prevail at a national level. Under these conditions, due to their differentiating impacts upon local economies, the subnational neocorporatisms of the 1970s explicitly counteracted the projects of national spatial equalization that were then still being pursued by most western European central governments.

In sum, the 1970s is best viewed as a transitional period characterized by intense interscalar struggles between political alliances concerned to preserve the nationalized institutional infrastructures of spatial Keynesianism and other, newly formed political coalitions concerned to introduce more decentralized frameworks of territorial development and urban governance. Although the new "projected spaces" sought by such modernizing coalitions remained relatively inchoate—generally they were limited to visions of a possible future geography rather than being actualized into stabilized institutional formations—such coalitions shared a broad commitment to the goal of endogenous growth and a more or less explicit rejection of nationally encompassing models of territorial development. In this sense, the proliferation of local and regional regulatory experiments, fueled by political coalitions oriented toward place-specific trajectories of socioeconomic development, articulated qualitatively new scale-making projects that markedly destabilized the nationalizing form of urban governance that had prevailed during the postwar "golden age." While central governments continued to promote such nationalizing, spatially redistributive agendas, the diffusion of this new bootstrap strategy during the 1970s appears, retrospectively, to have entailed a major de facto modification of the inherited institutional framework of spatial Keynesianism.

However, given the unstable character of these local regulatory experiments and scale-making projects, it would be an exaggeration to suggest that they contributed to the establishment of a new structured coherence for urban development. Rather, such bootstrap experiments are best understood as localized strategies of crisis management through which a new layer of state regulatory arrangements was superimposed upon the subnational political geographies that had prevailed during the period of urban managerialism. It was through the conflictual interaction of this newly emergent, subnational layer of state spatial regulation and the inherited national geographies of spatial Keynesianism that the broad contours of a new, rescaled landscape of state spatiality began to emerge as of the late 1970s. Insofar as the projects of endogenous growth of the 1970s entailed a clear divergence from the nationally equalizing, redistributive agendas of spatial Keynesianism, they established a significant politico-institutional opening for the more radical rescalings of urban governance and state spatiality that occurred during the subsequent decade.

The 1980s: Urban Entrepreneurialism and the Rise of Glocalization Strategies

The crisis of the Fordist developmental model intensified during the 1980s, leading to a new phase of industrial transformation, territorial reconfiguration, and state spatial restructuring throughout western Europe. The strategies of crisis management introduced during the 1970s had neither restored the conditions for a new growth cycle nor successfully resolved the deepening problems of economic stagnation, rising unemployment, and industrial decline within major western European cities and regions. Consequently, during the course of the 1980s, most European national governments abandoned traditional Keynesian macroeconomic policies in favor of monetarism: a competitive balance of payments subsequently replaced full employment as the overarching goal of monetary and fiscal policy. By the late 1980s, neoliberal political agendas such as welfare state retrenchment, trade liberalization, privatization, and deregulation had been adopted not only in the United Kingdom under Thatcher and in West Germany under Kohl, but also, in modified or hybrid forms, in many traditionally social democratic or social/Christian-democratic countries such as the Netherlands, Belgium, France, Spain, Denmark, and Sweden (Rhodes 1995).

This geopolitical sea change resulted in the imposition of additional fiscal constraints upon most municipal and metropolitan governments, whose revenues had already been significantly reduced during the preceding decade. Under these circumstances, political support for large-scale strategic planning projects waned and welfare state bureaucracies were increasingly dismantled, downsized, or restructured, not least at metropolitan and municipal levels. In the wake of these political realignments, major metropolitan institutions such as the Greater London Council and the Rijnmond in Rotterdam were summarily abolished during the mid-1980s. Elsewhere within western Europe, metropolitan institutions were formally preserved but significantly weakened in practice due to centrally imposed budgetary pressures and enhanced competition between city cores and suburban peripheries for capital investment and state subsidies (Barlow 1991). The fiscal squeeze upon public expenditure in cities and regions and the dissolution or weakening of metropolitan governance were thus among the major localized expressions of the processes of national welfare state retrenchment that began to unfold throughout western Europe during the 1980s. As of this decade, the national preconditions for municipal Keynesianism were being systematically eroded as local and metropolitan governments were increasingly forced to fend for themselves in securing a fiscal base for their regulatory activities (Mayer 1992).

During this same period, a new mosaic of urban and regional development began to crystallize throughout the western European city-system. As numerous observers have indicated, during the course of the 1980s a "self-reinforcing polarization of high-level activities in well-resourced and well-connected nodes" occurred throughout western Europe (Dunford and Perrons 1994, 172–73). Following the consolidation of the Single European Market and the launching of the euro during the 1990s, these polarizing tendencies were further entrenched both on national and on European scales. Consequently, on the European scale, economic activities were increasingly concentrated within a "vital axis" stretching from the South East of England, Brussels, and the Dutch Randstad through the German Rhinelands southward to Zürich and the northern Italian Industrial Triangle surrounding Milan (Dunford and Perrons 1994). In a now-famous report prepared for the French spatial planning agency DATAR prior to the consolidation of the Single European Market, Brunet (1989) famously described this core European urban zone as a "blue banana" whose strategic importance would be still further enhanced as geoeconomic and European economic integration proceeded.

Notably, Brunet's widely discussed representation of western Europe's urbanized boom zone represented a nearly exact inversion of the geography of development zones that had been promoted during the era of spatial Keynesianism. In stark contrast to the notions of cumulative causation upon which earlier spatial and regional policies had been based, in which the spatial diffusion of growth potentials was seen to benefit both cores and peripheries, Brunet's model implied that winning cities and regions would form a powerful, tightly interlinked, and relatively autonomous urban network dominated by advanced infrastructural facilities and high-value-added activities, leaving other regions essentially to fend for themselves or risk being marginalized still further in the new geoeconomic context. As Brunet's model rather dramatically illustrated, the tumultuous economic transformations of the 1980s were causing the economic geography of spatial Keynesianism to be turned essentially inside out: as of this decade, growth was no longer being spread outward from developed urban cores into the underdeveloped peripheries of each national economy but was being systematically reconcentrated into the most powerful agglomerations situated within Europe-wide spatial divisions of labor.

The remarkably wide influence of Brunet's model of the European blue banana was symptomatic of a major, state-led reorientation of urban governance that began to unfold throughout western Europe as of the 1980s. As urban economic restructuring intensified in conjunction with processes of global and European integration, western European central governments began more explicitly to target major cities and city-regions as the locational keys to national economic competitiveness. In the "Europe of regions" — a catchphrase that became highly important in national policy discussions by the mid-1980s — cities were no longer seen merely as containers of declining industries and socioeconomic problems but were increasingly viewed as dynamic growth engines through which national prosperity could be ensured. This view of the city as an essential national economic asset became dominant in mainstream policy circles in the late 1980s, as national and local governments attempted to prepare for the introduction of the Single European Market (Leitner and Sheppard 1998).

As Mayer (1994) notes, three broad realignments of urban governance subsequently ensued: (1) local authorities were constrained to engage more extensively and proactively in local economic development projects, (2) local welfarist and collective consumption policies were marginalized or subordinated to production-oriented policies, and (3) new forms of local governance, such as public-private partnerships, became widely prevalent. By the late 1980s, this new, entrepreneurial form of urban governance had been diffused throughout the European city-system as fiscally enfeebled local and regional governments mobilized a range of economic development strategies in order to attract inward investment (Harvey 1989b; Harding 1997; Parkinson 1991). Although urban entrepreneurialism was articulated in a variety of political forms during this period — including neoliberal, social democratic, and centrist variants (Eisenschitz and Gough 1993) —

all arguably contributed directly or indirectly to a Europe-wide diffusion of "beggar-thy-neighbor" interlocality competition, in which local and regional governments struggled aggressively against one another to lure a limited supply of investments and jobs into their territories (Peck and Tickell 1994; Cheshire and Gordon 1996).

As in the 1970s, the reorientation of urban governance during the 1980s was initiated in part through bottom-up strategies by local political coalitions struggling to manage the disruptive consequences of economic restructuring through ad hoc, uncoordinated policy adjustments and institutional shifts. As we saw above, however, the local economic initiatives of the 1970s emerged in a politico-institutional context in which central governments remained broadly committed to the social-democratic project of promoting national spatial equalization and sociospatial redistribution. In stark contrast, the entrepreneurial urban strategies of the 1980s were articulated under conditions in which neoliberal policy orthodoxies had acquired an unprecedented influence, leading in turn to a marginalization or even abandonment of traditional national compensatory regional policies in most western European states. In this transformed political context, the goal of equalizing economic development capacities across the national territory was increasingly seen to be incompatible with the new priority of promoting place-specific locational assets and endogenous growth within cities and city-regions. Accordingly, in addition to their efforts to undercut traditional redistributive regional policy relays, national governments mobilized a number of institutional restructuring strategies during the course of the 1980s in order to establish a new, competitive infrastructure for urban governance within their territories:

1. Local governments were granted new revenue-raising powers and an increased level of authority in determining local tax rates and user fees, even as national fiscal transfers to subnational levels were diminished (Fox Przeworski 1986).
2. Responsibilities for planning, economic development, social services, and spatial planning were devolved downward to subnational (regional and local) governments (Harding 1997; Parkinson 1991). In a number of western European countries, local economic development projects became key focal points for such devolutionary initiatives. Although these trends were

most apparent in traditionally centralized states, such as France and Spain, various policies to enhance regional and local autonomy were enacted in less centralized European states as well. In each case, decentralization policies were seen as a means to "limit the considerable welfare demands of urban areas and to encourage lower-level authorities to assume responsibility for growth policies that might reduce welfare burdens" (Harding 1994, 370). Even in the United Kingdom, where major aspects of local governance were subjected to increasing central control under the Thatcher regime, the problem of local economic governance was among the key issues upon which the restructuring of intergovernmental relations was focused (Duncan and Goodwin 1989).

3. National spatial planning systems were redefined. Economic priorities such as promoting structural competitiveness superseded traditional welfarist, redistributive priorities such as equity and spatial equalization. Meanwhile, in many European countries, the most globally competitive urban regions and industrial districts frequently replaced the national economy as the privileged target for major spatial planning initiatives and infrastructural investments. Particularly prominent instances of this realignment include the reorientation of Dutch spatial planning toward the Randstad megalopolis in the mid-1980s, the French national government's promotion of "megapolization" in the Paris–Ile de France region during the late 1980s, the refocusing of Danish regional planning on the Copenhagen region during the early 1990s, the introduction of new national urban growth policies in Italy after 1991, and the adoption of a city-centric approach to spatial planning in postunification Germany.
4. National, regional, and local governments introduced new, territory- and place-specific institutions and policies—from enterprise zones, urban development corporations, and airport development agencies to training and enterprise councils, inward investment agencies, and development planning boards—designed to reconcentrate or enhance socioeconomic assets within cities. Such institutions have often been autonomous from local state institutions and controlled by unaccountable political and economic elites. The Docklands redevelopment project in London and the Dutch mainports policies in Amsterdam and Rotterdam represent particularly prominent instances in which western European national governments

have channeled substantial public subsidies into strategically located urban economic development projects.

5. The forms and functions of local states were systematically redefined. Whereas postwar western European local governments had been devoted primarily to various forms of welfare service delivery, these institutions were transformed during the 1980s into entrepreneurial agencies oriented above all toward the promotion of economic development within their jurisdictions. This transformation occurred in three basic forms. First, confronted with increasing budgetary constraints, local states privatized or contracted out numerous public services and attempted to modernize systems of public administration. Second, local states attempted to promote economic regeneration by seeking to acquire subsidies through national or European industrial and sectoral programs. Third, and most crucially, local states—often in conjunction with regional state governments in federal countries—introduced a range of new policies to promote local economic growth, including labor market programs, industrial policies, infrastructural investments, place-marketing initiatives, and property redevelopment campaigns (see Eisenschitz and Gough 1993; Jessop 1998; Harding 1997).

Taken together, the institutional realignments reviewed above entailed not only a major reconfiguration of urban governance but an equally fundamental transformation within inherited geographies of state spatiality throughout western Europe. In contrast to the standardized geographies of state space under Fordism, in which national states attempted to maintain minimum levels of service provision and infrastructure investment throughout the national territory, the establishment of an entrepreneurial, competitiveness-oriented institutional infrastructure for urban governance during the 1980s entailed an increasing differentiation and fragmentation of state regulatory activities on various spatial scales. On the one hand, the consolidation of entrepreneurial forms of urban governance was premised upon the establishment of new subnational layers of state and parastate institutions through which cities could be marketed as customized, unique, and highly competitive locations for strategic economic functions within global and European spatial divisions of labor. On the other hand, the devolutionary and decentralizing initiatives mentioned above fundamentally reconfigured entrenched intergovernmental hierarchies and scalar divisions of regulation, imposing powerful new pressures upon all subnational administrative units to fend for themselves within an increasingly uncertain geopolitical and geoeconomic environment. In the face of these combined institutional realignments, inherited local and supralocal frameworks for urban governance were thoroughly reconfigured: the ***nationalizing*** scale-making project that had underpinned the Fordist-Keynesian model of urban governance was now being widely superseded by what might be termed a ***glocalizing*** scale-making project, the central goal of which was to position strategic local spaces competitively within global or supranational circuits of capital accumulation (Brenner 2001). In contrast to postwar strategies of spatial Keynesianism, which had contributed to an alleviation of intranational uneven development, these new glocalization strategies actively *intensified* the latter, (1) by promoting a systematic reconcentration of industry and population within each national territory's most globally competitive locations, (2) by encouraging increasingly divergent, place-specific forms of economic governance, welfare provision, and territorial administration within different local and regional economies, and (3) by institutionalizing intensely competitive relations, whether for public subsidies or for private investments, among major subnational administrative units (see also Goodwin and Painter 1996). A stylized contrast between spatial Keynesianism and glocalization as configurations of state spatiality and as approaches to urban governance is provided in table 6.1.

As the preceding discussion indicates, glocalization strategies began to play a particularly essential role within western European national states in the early 1980s, in close conjunction with the imposition of entrepreneurial forms of urban governance and the diffusion of interlocality competition throughout the Eu-

TABLE 6.1 Spatial Keynesianism and Glocalization: Two State Spatial Strategies

	Strategies of Spatial Keynesianism	Strategies of Glocalization
Geoeconomic and geopolitical context	•Differentiation of global economic activity among distinct national economic systems under "embedded liberalism" and the Bretton Woods monetary system	•New global-local tensions: global economic integration proceeds in tandem with an increasing dependence of large corporations upon local and regional agglomeration economies
	•The polarization of the world system into two geopolitical blocs under the cold war	•The end of the cold war and the globalization of U.S. dominated neoliberalism

(*continued*)

TABLE 6.1 (*continued*)

	Strategies of Spatial Keynesianism	Strategies of Glocalization
Privileged spatial target(s)	•National economy	•Urban and regional economies
Major goals	•*Deconcentration* of population, industry, and infrastructure investment from major urban centers into suburban and rural peripheries •*Replication* of standardized economic assets and investments across the national territory •Establishment of a nationally *standardized* system of infrastructural facilities integrated throughout the national economy •Alleviation of uneven development within national economies: uneven spatial development is seen as a *limit* or *barrier* to industrial growth	•*Reconcentration* of population, industry, and infrastructure investment into strategic urban and regional economies •*Differentiation* of national economic space into highly specialized urban and regional economies •Promotion of *customized,* place- and region-specific forms of infrastructural investment oriented toward global and European economic flows •Intensification and instrumentalization of interspatial competition within and beyond national borders: uneven spatial development is seen as a *basis* for industrial growth
Spatiotemporality of economic development	•National developmentalism: development of the entire national economy as an integrated, self-enclosed territorial unit moving along a linear developmental trajectory	•Glocal developmentalism: fragmentation of national economic space into distinct urban and regional economies with their own place-specific locational assets, competitive advantages, and developmental trajectories
Dominant policy mechanisms	•Locational subsidies to firms •Local welfare policies and collective consumption investments •Redistributive regional policies •National spatial planning and public infrastructural investments	•Deregulation and welfare state retrenchment •Decentralization of social and economic policies and fiscal responsibilities •National urban policies and spatially selective investments in advanced infrastructures •Regional industrial policies •Local economic initiatives •Metropolitan institutional reform
Dominant ideological slogans	•"National development"; "national economy"; "progress" •"Balanced growth" or "*Ausgleich*" •"Balanced urbanization" •"Government"; "public goods"	•"Globalization"; "erosion of the state"; "deterritorialization" •"Cities in competition"; "*Standortkonkurrenz*" •"Europe of the regions"; "Europe of the cities" •"Endogenous development" •"Governance"; "lean management"

ropean urban system. However, insofar as glocalization strategies first emerged within formations of state spatiality that had been inherited from the Fordist-Keynesian era, their institutional consequences were necessarily highly uneven, varying considerably across western Europe according to (1) the nature of inherited patterns of state spatial organization and intergovernmental relations, (2) the specific restructuring strategies adopted by modernizing political alliances within each national context, and (3) the resilience of conservative/preservationist political alliances in attempting to defend the national social compromises associated with the Fordist-Keynesian order.

In general terms, therefore, the new, rescaled geographies of state spatiality and urban governance that crystallized during the second half of the 1980s can be conceived as expressions of a complex, path-dependent, and conflictual *interaction* between newly emergent glocalization strategies, with their predominant goal of promoting urban reconcentration and spatial differentiation, and the nationally configured regulatory geographies that had been inherited, albeit in a destabilized form, from the transitional period of the 1970s. Thus understood, the diffusion of glocalization strategies in western European states during the 1980s did not simply erase earlier geographies of state regulation but generated contextually specific spatial *rearticulations* of inherited and emergent state regulatory practices on a range of geographical scales.

The rescaled formation of national state spatiality that crystallized through these contested transformations may be provisionally characterized as a glocalizing competition state regime (GCSR)—*glocalizing* because it rests upon concerted national political strategies to position diverse subnational spaces (localities, cities, regions, industrial districts) within supranational (European or global) circuits of economic activity (Swyngedouw 2000); a *competition state* because it privileges the goals of structural competitiveness and flexibility over traditional welfarist priorities such as equity and redistribution (Jessop 2000); and a *regime* because it represents an unstable, uncoordinated, and continually evolving spatial mosaic of political strategies, institutional modifications, and regulatory experiments rather than a fully consolidated state form (Peck and Tickell 1994). Whether forged predominantly from below, through the initiatives of local and regional coalitions to manage economic restructuring, or from above, through central state strategies to enhance the competitiveness of local and regional economies, the institutional infrastructures of urban entrepreneurialism soon came to occupy pivotal regulatory positions within the newly crafted scalar architectures of GCSRs. It is in this specific sense that the rise of entrepreneurial forms of urban governance provided an important institutional mechanism for the rescaling of national state spaces during the 1980s.

■ The 1990s: The Ambiguous Resurgence of Metropolitan Regionalism

A number of commentators have emphasized the chronically unstable character of entrepreneurial approaches to urban governance (Leitner and Sheppard 1998). Because entrepreneurial urban strategies intensify social inequality, social exclusion, and uneven spatial development, they ultimately undermine the very socioterritorial conditions upon which sustainable capitalist growth is contingent (Peck and Tickell 1994). While entrepreneurial urban strategies may successfully unleash short-term bursts of economic growth within circumscribed local and regional economies, they generally fail to sustain such economic upswings beyond the medium term. These contradictions within entrepreneurial forms of urban governance have been exacerbated during the 1990s in conjunction with the diffusion of GCSRs throughout western Europe. As indicated, one of the major effects of glocalization strategies has been to enhance competitive pressures upon all subnational administrative units and thus to intensify uneven geographical development still further within each national territory. While these institutional realignments may temporarily benefit a select number of powerful, globally competitive urban regions, they inflict a logic of regulatory undercutting upon most local and regional economies, a trend that may seriously downgrade national economic performance in the medium and long term. At the same time, the increasing geographical differentiation of state regulatory activities induced through glocalization strategies is "as much a hindrance as a help to regulation" (Goodwin and Painter 1996, 646). For, in the absence of institutional mechanisms of metagovernance capable of coordinating subnational regulatory initiatives and competitive strategies, these ongoing rescaling processes may undermine the state's organizational coherence and functional unity, leading in turn to serious governance failures and legitimation deficits (Jessop 1998).

The contradictory tendencies unleashed during the last two decades of state rescaling have had important ramifications for the evolutionary trajectories of GCSRs.

Faced with the pervasive regulatory deficits of their own predominant strategies of economic regulation, many glocalizing competition state regimes have been constrained, during the course of the 1990s, to engage in various forms of institutional restructuring through which to manage the disruptive, dysfunctional socioeconomic consequences of unfettered interlocality competition. Thus, whereas the rescaling of urban governance during the 1970s and 1980s was animated primarily through strategies to manage economic crisis and to rejuvenate industrial growth within major local and regional economies, the rescaling projects of the 1990s have been mediated increasingly through strategies designed to manage the pervasive governance failures associated with the previous round of regulatory restructuring and state rescaling. In this manner, during the 1990s, a variety of political responses to what Offe (1984) once termed "the crisis of crisis-management" have been superimposed upon the local economic initiatives and crisis-management strategies that were initially mobilized in western European cities and states following the economic downturn of the 1970s. Political strategies designed to manage this crisis of crisis management have arguably played an essential role in reshaping the institutional and geographical architectures of GCSRs since the early 1990s, when the contradictions of first-wave glocalization strategies and entrepreneurial urban strategies became widely apparent throughout western Europe. During this period, glocalization strategies began increasingly to encompass not only entrepreneurial approaches to urban development but also a variety of local and supralocal "flanking mechanisms and supporting measures" (Jessop 1998, 97–98) intended to manage the diverse tensions, conflicts, and contradictions generated by such policies both within and beyond localities. Although these newly emergent strategies of crisis management have not prevented the aforementioned contradictions from being generated, they have nonetheless entailed the establishment of diverse institutional mechanisms through which the most disruptive political-economic consequences of such contradictions may be monitored and managed.

It is in this context, I would argue, that the widespread proliferation of new *regionally* focused scale-making projects during the 1990s must be understood (Heinz 2000; Keating 1997; MacLeod 2000). As indicated, the first wave of glocalization strategies during the 1980s focused predominantly upon the downscaling and decentralization of formerly nationalized administrative capacities and regulatory arrangements toward local tiers of state power. It was under these conditions that many of the metropolitan institutional forms that had been inherited from the Fordist-Keynesian period were abolished or downgraded. More recently, however, the metropolitan and regional scales have become strategically important sites for a new round of regulatory experiments and institutional shifts throughout western Europe. From experiments in metropolitan institutional reform and decentralized regional industrial policy in Germany, Italy, France, and the Netherlands to the Blairite project of establishing a patchwork of Regional Development Agencies (RDAs) throughout the United Kingdom, these developments have led many commentators to predict that a "new regionalism" is superseding both the geographies of spatial Keynesianism *and* the forms of urban entrepreneurialism that emerged immediately following the initial crisis of North Atlantic Fordism (Cooke and Morgan 1995). Against such arguments, however, the preceding discussion points toward a crisis-theoretical interpretation of these initiatives as an important *evolutionary modification* of GCSRs in conjunction with their own immanent contradictions. Although the politico-institutional content of contemporary regionalization strategies continues to be an object of intense contestation, they have been articulated thus far in at least two basic forms.

On the one hand, regionally focused strategies of state rescaling have frequently attempted to transpose entrepreneurial approaches to local economic policy upward onto a regional scale, leading in turn to a further intensification of uneven spatial development throughout each national territory. In this scenario, the contradictions of urban entrepreneurialism are to be resolved through the integration of local economies into larger, regionally configured territorial units, which are in turn to be promoted as integrated, unified, and competitive locations for global and European capital investment. In this approach to regional state rescaling, the scalar configuration of GCSRs is being modified in order to emphasize regions rather than localities; yet the basic politics of spatial reconcentration, unfettered interspatial competition, and intensified uneven development are being maintained unchecked. On the other hand, many contemporary strategies of regionalization have attempted provisionally to countervail the dynamics of unfettered interlocality competition by promoting selected forms of social redistribution, social cohesion, and spatial equalization *within* strategic regional institutional spaces. Although such initiatives generally do not significantly undermine uneven development between regions, they can nonetheless be viewed as efforts to modify some of the most disruptive local and regional impacts of the glocalization strategies that prevailed during the 1980s, particularly in the context of intensifying city-suburban conflicts. Indeed, this aspect of regional state rescaling may be viewed as an attempt to reintroduce a downscaled form of spatial Keynesianism *within* the subnational regulatory architecture of glocalizing states: the priority of promoting equalized, balanced growth is thus being promoted on a regional scale, within delimited subnational zones, rather than throughout the entire national territory.

Which mixture of these opposed glocalization strategies may prevail within a given national, regional, or local institutional environment hinges upon intense sociopolitical struggles in which diverse social forces strive to influence the geography of state regulatory activities and private investment toward particular political ends. Nonetheless, both of these new, rescaled forms of crisis management appear to represent significant evolutionary modifications within the GCSRs that were consolidated during the 1980s. In this newest scale-making project of the 1990s, the twin priorities of economic competitiveness and crisis management are juxtaposed uneasily in an unstable, continually shifting institutional matrix for urban and regional governance. While there is currently little evidence to suggest that either of these regionalized glocalization strategies will engender sustainable forms of economic regeneration in the medium term, they are nonetheless likely to continue to intensify the geographical differentiation of state space and the uneven development of capital throughout western Europe, leading in turn to a new historical constellation of contradictions, crisis tendencies, institutional responses, and sociopolitical conflicts in major cities and city-regions.

Conclusion: Cities, Glocalization Strategies, and the New Landscape of Regulation

This chapter has attempted to demonstrate how the rise of entrepreneurial approaches to urban governance in western European cities has been intertwined with a broader rescaling of national state spaces following the dismantling of spatial Keynesianism in the late 1970s. The concept of the glocalizing competition state regime (GCSR), as summarized here, is intended to provide a theoretical basis on which to grasp the tangled new layerings of state spatiality that have been produced through these conflictual rescaling processes. Within the newly emergent, glocalized configuration of state spatiality, national governments have not simply transferred power downward but have attempted to institutionalize competitive relations between major subnational administrative units as a means to position local and regional economies strategically within supranational (European and global) circuits of capital. In this sense, even in the midst of the pervasive relativization of scales that has unsettled traditional national regulatory arrangements, central governments have attempted to retain control over major subnational political-economic spaces by integrating them within operationally rescaled, but still nationally coordinated, accumulation strategies. I have suggested, in this context, that the contradictions unleashed within GCSRs provide an important impetus for their further political, institutional, and geographical evolution, in large part through the production of new (regional) scales of state spatial regulation in which crisis-management strategies may be mobilized. It is in the context of these emergent, scale-sensitive forms of crisis management, I believe, that the ongoing shift from a "new localism" to a "new regionalism" in many western European states (Deas and Ward 2000; Keating 1997; MacLeod 2000) must be understood.

As this analysis has indicated, the rescaled configurations of state spatiality that have been consolidated during the last two decades of capitalist restructuring have systematically undermined the nationalized forms of social and spatial justice that were established in postwar Europe in the wake of many decades of political struggle. This newly imposed, rescaled landscape of market-oriented political regulation has generated new forms of sociospatial inequality and political conflict that significantly limit the choices available to progressive forces throughout Europe both on local and on national scales: for, in the current geoeconomic climate, the project of promoting territorial equalization within national or subnational political units is generally seen as a luxury of a bygone era that can no longer be afforded in an age of lean management and fiscal austerity. Yet, even as these rescaling processes appear to close off some avenues of political regulation and democratic control, they may also establish new possibilities for sociospatial redistribution and progressive, radical-democratic political mobilization on other spatial scales. The processes of European integration and enlargement have to date been dominated by orthodox neoliberal agendas that reinforce and even intensify the entrepreneurial politics of interspatial competition described in the preceding discussion (Agnew 2001). Nonetheless, the supranational institutional arenas associated with the EU may still provide a potentially powerful political mechanism through which progressive forces might once again mobilize social and spatial programs designed to alleviate inequality, uneven development, and unfettered market competition, this time on a still broader spatial scale than was ever thought possible during the era of high Fordism (Dunford and Perrons 1994).

It remains to be seen whether the contemporary dynamics of glocalization and state rescaling will continue to be steered toward the perpetuation of neoliberal geographies of uneven spatial development based upon intensifying inequality and social exclusion or whether—perhaps through the very contradictions and conflicts they unleash—they might be rechanneled to forge a negotiated political compromise on a European scale based upon substantive social and political priorities such as democracy, equality, diversity, and increased free time. Precisely because the institutional and scalar framework of European state space is in a period of profound flux, its future can be decided only through ongoing sociopolitical struggles, at once on local, national, and supranational scales, to rework the geographies of regulation and political mobilization. Under conditions such as these, the spatiality of state power has become the very object and stake of such struggles rather than a mere arena in which they unfold. A major intellectual and political task for state theorists, political sociologists, and political geographers in the current era is to develop new theoretical frameworks through which to grasp the intensely contested processes of state spatial restructuring that have been provoked by such struggles.

NOTE

1. The reorientation of urban governance from the managerial, welfarist mode of the Fordist-Keynesian period to an entrepreneurial, competitiveness-oriented framework during the post-1970s period has been analyzed extensively during the last decade in the vast literatures on urban political economy (see, e.g., Harvey 1989b; Hall and Hubbard 1998). However, with a few notable exceptions (e.g., Jessop 1998; Macleod and Goodwin 1999), the mediating links between this restructuring of urban governance and broader transformations of state spatiality have not been examined or theorized.

PART III

Europe as Experience

SEVEN

Ways of Seeing European Integration

Germany, Great Britain, and Spain

JUAN DÍEZ-MEDRANO

Public support for European political integration is significantly lower in Great Britain than in Spain or Germany. Because of this, the British are often portrayed as deviant, as spoil-sports who are unable to adapt to a changing world in which states have become too small. But are they really? Shouldn't we ask instead how it is that, comparatively speaking, the populations of countries like Spain and Germany so easily accept the withering away of the nation-state? It takes a somewhat structuralist-functionalist and ahistorical view of social change to expect that sweeping changes in the economy, such as globalization, will be automatically followed by appropriate political and mental adjustments toward the development of at least regional, if not global, forms of political organization.

Rapid and extensive social change inevitably generates resistance. Peasants did not happily turn into industrial workers, local elites did not happily surrender their power to the emergent modern state, and the "nationalization of the masses," to borrow a concept from the late George Mosse (1975), did not just happen. The latter, in particular, was a long and often bloody process, in which the state itself was a leading player. To foresee in 1918 that the same century that began with the apogee of nationalism would come to a close under unequivocal signs of an uncoupling of national and political identities would have required a good deal of imagination. Therefore, explaining some people's apparent willingness to strengthen European political institutions is as important as explaining other people's reluctance to do so.

In this chapter I provide an overview of how ordinary citizens, local elites, and public intellectuals in Germany, Great Britain, and Spain conceive of European integration, and I analyze the relationships that exist between these conceptions. This focus on cognitive frames about European integration is indispensable for a better understanding of how attitudes toward European integration develop and how they affect the politics of European unification. People's images of European integration and of the European Union, and not the objective historical factors that caused them, are after all the raw material with which national and European Union political elites have to work in their efforts to reinforce or change the way people feel about these political developments.[1]

■ Cultural Repertoires, Narratives, and Collective Memory

In contrast to extant explanations of individual variation in support for European integration that focus on factors such as cost-benefit calculations (Hewstone 1986; Gabel 1998; Eichenberg and Dalton 1993) and levels of cognitive mobilization and postmaterialist values (Deutsch 1953; Inglehart 1977) but largely fail to explain intra-European variation in support for European integration, my approach stresses the roles of cultural repertoires and elite-mass diffusion processes. It thus bears a resemblance to another tradition, more historical and cultural, that has also been popular among social scientists when discussing European integration (Barzini 1983; Schultz 1984; W. Wallace 1986; Vernet 1992; Torregrosa 1993; Mommsen 1994; Álvarez-Miranda 1996). In comparison with the purely cultural and historical literature, my investigation relies on a more systematic collection and analysis of empirical information and on a thorough examination of elite-mass diffusion processes.

The literature on cultural repertoires includes work on social representations (Moscovici 1984), cultural repertoires (Swidler 1986), narratives (W. Wallace 1986; Sewell 1991), and collective memory (Halbwachs [1925] 1975; Fentress and Wickham 1992). This literature stresses that people approach problems by drawing, consciously or unconsciously, on habits, stories, skills, rituals, and worldviews. My focus on cultural repertoires coincides with recent interest in cross-national contrasts between cultural repertoires in the field of comparative cultural sociology (Lamont and Thévénot 2000). This emerging tradition attempts to avoid essentialist views of national cultures, by stressing that cultural repertoires are often shared by different nation-states but in different proportions. In this chapter I demonstrate that both elites and masses, when arguing for or against European integration, call upon specific and contrasting narratives about the meaning and expected consequences of European integration and of membership in the European Community (e.g., the British emphasize the democratic deficit and the lack of accountability in the European Union). International variation in support for European integration results not so much from the contrasts between singular national narratives but rather from contrasts in the relative salience of specific narratives across countries.

I draw also on elite-mass diffusion models. There is indeed a long tradition in the social sciences concerned with the role of elites in shaping people's values and beliefs (Deutsch 1968; Rosenau 1961; Petersen 1972; Dalton and Duval 1981).[2] In general, these models suggest that elites influence opinion leaders, primarily through articles and speeches published in the quality press; in turn, these opinion leaders influence politically aware citizens through word of mouth. Elite-mass diffusion models also assume that those who are not interested in politics get their information primarily from TV news programs and from sketchy news reports in local, popular, or tabloid newspapers. These other sources of information generally provide factual information but are also often framed in a way that amplifies and simplifies the ideas and messages produced by the elites and published in the quality press (Bröder 1976).

A goal of this chapter is to analyze the manner in which public intellectuals have reflected on European integration and to determine whether there is correspondence between their views of European integration and those of the public at large. To capture this public intellectual discourse, my analysis focuses on the *quality* press rather than on the popular press or on TV, even though the quality press is read by a much smaller segment of the population. Because of my focus on elite discourse, I do not examine the complex process through which ideas get to the public or the effects of the media on public opinion. My analysis demonstrates the stability of the German, British, and Spanish frames about European integration since the 1950s and simultaneously provides information that is consistent with theories that posit an influence of elites on the general public. It shows that there is a great similarity between comments made about European integration and membership in European institutions by public intellectuals and the rest of the citizens. Despite the consistency of the findings with elite-mass diffusion models, the results presented here should not be interpreted as proof of the existence of a decisive causal link between elite and mass frames. It is indeed possible that citizens in general, whatever their social position, draw on the same cultural materials to develop their views of European integration. Furthermore, the results of my analysis suggest that citizens are selective, that is, that they are as much influenced by elite opinions as by factual reports about the routine functioning of the European institutions.

■ Ordinary Citizens, Local Elites, and European Integration

Between February 1996 and November 1997, I conducted interviews with ordinary citizens and local elites in two different cities in each of the three countries included in my investigation. The cities were Weststadt, Oststadt, Quijotón, Catadell, Engleton, and Scotsburg.[3] The cities were selected to represent regions with distinctive national identities in Germany, Spain, and Great Britain, as well as the social and economic structures corresponding to the regions in which the cities were located. I used statistical data provided by Eurostat for the European Union's regions. The three indicators that I relied upon were (1) the distribution by economic sector of the active population, (2) mean income per capita, and (3) unemployment rates. I ended up with a west German and an east German city for Germany (Weststadt and Oststadt), a Castilian and a Catalan city for Spain (Quijotón and Catadell), and an English and a Scottish city for Great Britain (Engleton and Scotsburg).

I relied on systematic sampling of entries in the cities' telephone books to select my group of respondents. My target sample consisted of approximately 18 respondents, which I selected so that I would obtain a similar number of respondents for various age and education combinations. My entire sample consisted of 160 interview subjects. I defined three age groups (under 30, 31 to 50, and over 50) and two education groups (less than high school and more than high school). The end result was a grid composed of six cells, with three interviews per cell.[4]

As part of my fieldwork in each of the six cities, I also conducted interviews with members of the local elite. The selection was guided by an attempt to represent all groups likely to influence public opinion on the topic of European integration. The respondents included leading members of the local branches of the main political parties in the region where the interviews took place, the editor of the local paper, and representatives of the local chamber of commerce and industry, of the main workers' union, and of one of the town's housewives' or women's associations. In Oststadt, I also interviewed members of the local association for the unemployed.

Each interview was taped and lasted for about an hour. The questionnaire was based on a series of general closed- and open-ended questions related to people's attitudes toward European integration, preceded by warm-up questions concerning the respondents' degree of interest in politics, degree of media exposure, and perception of the main problems affecting the country. The questionnaire ended with some questions about the respondents' identification with their region, their country, and Europe. I personally transcribed, coded, and analyzed the interviews.

In the discussion that follows, I show that cognitive frames with respect to the European integration in Germany, Great Britain, and Spain overlap but also contain clearly distinct elements. These cognitive frames are by and large shared by all segments of each country's population, whether ordinary citizens or members of the local elite, informed or uninformed, young or old, men or women, favorable to European integration or against European integration. Therefore, I will devote little space to analyzing such socio-demographic distinctions.

■ Modes of Argumentation

Thinking about European integration is not a popular leisure activity among ordinary citizens who hear about Europe once in a while in the news, register the information, and then forget it. An increasing number of people are beginning to confront the European Union directly, however. The European Monetary Union is the most striking example, since it is the first political measure in the process of European integration whose impact will be immediately felt by everybody who resides in the European Union. Apart from this, the European Union makes itself visible in many ways: to farmers who must adjust what they produce and in what quantities to meet the requirements of the Common Agricultural Policy (CAP); to students, who ponder whether to study in another EU country and apply for an Erasmus Fellowship; and, more generally, to ordinary citizens who travel to other EU countries and rejoice at not having to carry a passport. The number of those who at some point or another confront the EU during their daily activities is already significant, but except for the few whose livelihood is highly dependent on what the European institutions decide, the majority of the population pays at least as little attention to EU affairs as it pays to TV commercials.

Repetition pays, however, and just as people eventually respond to commercials, people have accumulated some knowledge about the EU over the years, enough to develop a certain image of the EU: a sketch of what it represents, what it promises, how it works, and what the costs and benefits of membership are. When one puts together all these carefully assembled arguments, sketch over sketch, the contours of people's images of the EU become clearer, and differences and similarities between cities and countries emerge. Participants in this project provided a wide variety of justifications for their answers on issues related to European integration, spanning more than 130 favorable arguments and more than 100 unfavorable ones. In fact, these are conservative estimates, since similar answers were coded together into broader categories. For instance, respondents who justified their support for European integration by saying, "States are too small," and those who said, "Countries are too small," were coded under the same category.

Although there is a great deal of diversity in the way people think about European integration, the pages that follow reveal that the respondents' views were, in a most Weberian fashion, predominantly informed by considerations of economic interest, status, and power, and also by considerations of good and democratic governance and geopolitical stability. These considerations were in turn guided by past and current individual and collective experience and by expectations about what the future will bring. The present shapes people's views of European integration through their lived experiences and the information they get from local and national sources. The past conjures up images of better or worse times, to which the present and the future can be compared. It also makes us conscious of what could happen and mobilizes us to find strategies to avoid it or bring it back. These strategies are in turn informed by our causal reading of past events. Last but not least, the past reminds us of what we are or once were. Finally, people's images of European integration and the European Union are informed by what they promise for the future. These promises are fundamentally shaped by the messages one gets through the media, but they are also increasingly filtered through people's experiences and interpretations of the effects that European integration has had up to the present.

Participants in this project provided a wide variety of justifications for their answers on issues related to European integration. The most important ones are presented in table 7.1. Respondents in all three countries mentioned some themes. They constitute the internationally shared core of people's cognitive frames. On the positive side, the themes that were mentioned most often were the advantages associated with the common market, the belief that states have become too small to compete in a globalized world, and the removal of passport controls between the countries that belong to the European Union. On the negative side, most of the comments were focused on governance aspects of the European Union. The European Union was portrayed as opaque, distant, inefficient, inadequate to respond to the task of unification, unaccountable, paralyzed by national egoisms, eroded by corruption, and obsessed with regulating every little aspect of people's lives.

Beyond these commonalities, there were very clear differences between the countries. I will begin with Germany and focus on the most significant results.

■ Germany: Labor Competition, Democratic Deficit, and National Socialism

Despite their faith in the common market and its economic advantages, the German population's optimistic approach to European unification was clouded by their anxiety about the effects of differences in wages and general social benefits across the European Union. High unemployment rates, the completion of the European Union's internal market for goods, labor, and capital, and the fact that German wages and social benefits are higher than those in other European Union countries explain this anxiety. The problem of unemployment in Germany has become inextricably linked to the problem of wage and social benefit inequality across the European Union. The arguments I heard in the interviews reveal that restrictions on the functioning of the competitive mechanisms of the internal market as well as immigration restrictions, increased taxation of companies that relocate abroad, and more stringent controls on foreign labor competition within the national market are the solutions that people are contemplating.

If we now focus on regional differences within Germany, we see that east German and west German respondents (Oststadters and Weststadters, respectively) shared a very similar cognitive frame about European integration. In one

TABLE 7.1 Themes Mentioned to Justify Positive or Negative Comments about European Integration or the European Union, by City

Number of Respondents Mentioning a Particular Theme

	Germany		Spain		Great Britain		
	Weststadt	Oststadt	Quijotón	Catadell	Engleton	Scotsburg	Total
Common market	23	22	19	20	23	22	129
States too small	12	14	16	14	13	10	79
Removal of barriers	13	10	11	8	13	12	67
Governance	12	10	9	9	14	16	70
Free movt. and competition	10	10	1				21
Democratic deficit	11	6	4		5	1	27
Lessons of WWII	17	2					19
Peace	11	9	4	1	9	8	42
Understanding	16	5	9	6	11	6	53
Modernization	1		8	9	3	3	24
CAP	1	4	15	12	7	8	47
Structural/regional funds	1	1	12	10	6	9	39
Against isolation		1	7	5	2		15
Lack of voice		2	4	8	3	3	20
Sovereignty and identity	2	1	2	2	9	9	25
Social benefits	1	2	4	4	5	8	24
Total respondents	27	29	27	24	27	26	160

NOTES: Understanding: Contributes to better understanding between peoples and cultures; Common market: The common market is economically beneficial; CAP: The Common Agricultural Policy is a bad policy; Democratic deficit: European institutions suffer from a democratic deficit; States too small: States are too small to face economic or military challenges; Governance: The governance of European institutions is poor; Against isolation: Membership of this country is necessary to break the country's isolation, isolation would be disadvantageous for the country; Lack of voice: The country's voice is not taken into account within European institutions; Modernization: The country will modernize as a member of European institutions; Removal of barriers: The removal of barriers to the movement of people is a good thing; Social benefits: The country's social benefits will increase as a result of membership in the European institutions; Sovereignty and identity: Membership in the European Union has or will have a negative effect on sovereignty and identity; Lessons of WWII: Membership in the European institutions will reduce misgivings toward the country; Peace: Will contribute to peace; Free movt. and competition: Free movement of workers will mean competition from foreign workers; Structural/regional funds: The structural and regional funds of the European institutions are a good thing.

critical respect, however, their collective representations of European integration differed radically. This concerns whether or not they mentioned Germany's role in the Second World War as an explanation for why Germany must contribute to European integration and be a member of the European Union.

There were three slightly different versions of this argument. The first one was that Germany has a moral obligation to cooperate, so as to compensate for Germany's behavior in the Second World War. The second one was that Germany should be a member of the European Union to protect itself against its own demons, that is, its tendency to become hegemonic and aggressive whenever it feels strong. The third and most frequently heard one was that Germany should be a member of the European Union to reassure the world about its peaceful intentions: to prove to other countries that Germany does not harbor intentions to channel its economic power into efforts toward political and military hegemony over Europe. This argument was sometimes, but not always, followed by resignation or frustration at what respondents considered to be an unjust situation.

At any rate, the discussion above shows that Germany's role in the Second World War plays a significant role in the way west Germans think about European unification, but a role that is mediated by their perceptions of the current social and political situation in Germany. The finding that this theme played practically no role in the way Oststadt's respondents discussed the European Union is quite puzzling, since studies conducted after reunification do not show substantial differences in the way west and east Germans remember the Second World War or evaluate Germany's role in it when asked about it (Horst 1996). One plausible explanation for the contrast between west and east Germans, which is consistent with contrasts in the contents of high school contemporary history textbooks, for example,[5] is that Germany's responsibility in the Second World War, although remembered and acknowledged, is a much less salient theme in east Germany than it is in west Germany and was not linked with European integration during the Communist period.

■ Spain: Modernization, Status, and the Franco Regime

Moving now to Spain, we see that modernization, status, and opposition to isolationism were the distinctive themes that framed the Spanish respondents' thinking on the topic of European integration. The first theme encompasses statements by which respondents justified Spain's membership in the European Union with the belief that membership will contribute to the modernization of the country. Tied to this belief is the respondents' sensitivity to the European Union's economic policies. Thus, Spanish respondents distinguished themselves from German respondents, but not so much from British respondents, by their higher propensity to mention the Common Agricultural Policy and aid received from the European Union as sources of dissatisfaction and satisfaction, respectively.

The themes of status and modernization intersect in the relatively high frequency with which Spanish respondents noted with frustration that Spain's voice is not heard in the European Union. This was particularly true in Catadell, where one out of three respondents referred to this problem. The comments I heard reflected the high salience that the European Union's expected contribution to the modernization of Spain has in the Spaniards' *conscience collective*. Also, however, it reflected a general feeling that Spain is being taken advantage of and that its opinions are not taken into account. This, in turn, was frequently interpreted as a sign that Spain is not yet treated as an equal but, rather, as non-European. The European Union has therefore become a mirror for Spain. EU decisions that affect Spain are treated either as major victories or as major setbacks, as recognition of Spain's worth or as major affronts.

The third and final theme that emerged from the Spanish respondents' discussions of European integration was the desire to break with Spain's tradition of isolation. This desire ties the attitudes of Spanish respondents to their memory of the period of Franco's dictatorship and serves as the connecting point for people's preoccupation with modernization and status. In this sense, to the extent that it reminds people of a long period of economic stagnation and international isolation, the dictatorship of General Franco (1939–75) has the same cultural and political significance for Spaniards as National Socialism (1933–45) has for the Germans, or the British empire and victory in the Second World War for the British.

■ Great Britain: Sovereignty, Identity, and the Welfare State

The most distinctive aspect of the British respondents' way of thinking about European integration was the significant role played by sovereignty and, more significantly, national identity. Many of the comments referred to people's fear of losing the nation's identity, culture, and way of life if political integration moves forward. This fear was often grounded on the respondents' belief that Great Britain has little clout in the decision-making process of the European Union and would therefore be in no position to defend its national identity and culture if the European Union increased its supranational character. They see the European Union as an extraneous bureaucracy that gives orders to Great Britain, a sort of dictatorship, instead of an institution in which Great Britain is represented and plays a role.

In discussing the topic of identity, British respondents stressed the differences between British culture and the cultures of other European countries.

Sometimes, this emphasis on cultural differences led to a Herderian glorification of the advantages of cultural variety. On other occasions, however, comments about the irreconcilable cultural differences between European countries were followed by a nationalistic glorification of the British culture.

National Self-perceptions, Collective Memory, and European Integration

The examination of the themes that people raised when discussing European integration gives us privileged access to the mental sketches that frame their thinking about this topic. As I have shown, these sketches were similar in fundamental ways (e.g., the common market, globalization, governance) but also differed across countries in ways that are no less significant (e.g., the role played by the country in the Second World War, modernization, identity, and sovereignty).

Beyond this, the analysis of people's answers during the interviews is also useful for the development of an explanation of people's response to European integration because it reveals the significance of national self-perceptions and collective memory. I have coded these comments in order to detect regularities that may smooth the transition from an intersubjective to an objective and historically grounded explanation of differences in attitudes toward European integration in Great Britain, Germany, and Spain.[6]

Among German respondents, for instance, most of the comments related to national self-perception were of a positive nature. They saw their country as rich and powerful (about 1 in 5 respondents), as having better legislation and policies than other countries (about 1 in 10 respondents), and as permeated by a strong work ethic. Among Spanish respondents, most of the comments about national self-perception were negative. Spain was seen as a poor and powerless country (about 1 in 4 respondents), a country with bad legislation and policies, ruled by incompetent elites, and where people tend to lack discipline. Among Britons, views were divided. Some respondents viewed the country as rich and powerful and thought that a main feature of its character was fairness. Other respondents, however, portrayed the country as being poor and powerless or as a declining power. Beyond these positive and negative assessments, respondents tended to portray the British population as anti-European, isolationist or parochial, and xenophobic or nationalist. Sometimes such comments were meant to be critical, but on other occasions they were simply stated as fact. These divisions among the British respondents, and the fact that they made many more comments about their country, their people, and even about other countries than did Spanish or German respondents, convey that there is less consensus about how to characterize the country in Great Britain than there is in Spain or Germany. Moreover, the high frequency with which respondents in Great Britain critically referred to British nationalism and xenophobia indicates a significant split in the population with regard to national identity, ethnic relations, and attitudes toward the European Union.

During the interviews respondents also referred, directly or indirectly, to key moments of their history when they pronounced themselves in favor of or against European integration. In order to systematically assess the salience of particular historical events in people's discussions of European integration, I coded all historical events mentioned during the interviews. In total, respondents invoked more than fifty different historical events or periods. The distribution of mentioned events shows that west German respondents articulated their reflections around the Second World War; east German respondents articulated theirs around the reunification process and Communism; Spanish respondents articulated theirs around the Franco period; and British respondents articulated theirs around the Allied victory in the Second World War, the British empire, and the commonwealth.

The images that respondents have of the European Union must therefore be interpreted with reference to the national cultures within which they have developed. The content analysis of the interviews I conducted in Germany, Great Britain, and Spain reveals that national self-perceptions and historical memory are the most relevant dimensions of this cultural context. The impression conveyed by my respondents in Weststadt is that they were very proud of Germany's economic and social achievements during the postwar period. In terms of its work ethic, standards of living, and social benefits, they saw Germany as superior to the rest of Europe. Therefore, they said they would like the European Union to be shaped along the German model rather than give up some of their achievements. This attitude becomes especially salient when it comes to facing and addressing current high unemployment levels. One could interpret the current sensitivity to competition from foreign workers and pressure to shape the European Union along the German model as resulting from this combination of pride in past achievements and fear that Germany's economic and social model may be under strain. The respondents tended to blame others rather than to question the German model. However, west German respondents were well aware that Germany's image abroad is still colored by memories of the Second World War and the Holocaust; accordingly, they expressed commitment to European unification and a desire to solve Germany's current economic predicament within the framework of the European Union. Hence the solution that I heard most often, which was the extension of Germany's social model to the rest of the countries of the European Union.

The east German respondents I interviewed in Oststadt resembled their west German counterparts only in the degree of confidence they expressed in the superiority of the German model and the German character. Although they considered themselves to be under tremendous economic strain, they attributed this not to failings of the German social and economic model, which they welcomed, but rather to the way reunification was planned and executed. On the national level, they felt abandoned and somewhat betrayed by their western counterparts, but from a European perspective they saw themselves as part of a prosperous country and therefore well above the rest of Europe. Because of unemployment rates that are much higher than they are in west Germany, their reaction when it came to foreign workers was quite similar to that in west Germany: one should prevent foreign workers from coming, and if they do come, one should make sure that they work under the same conditions as the autochthonous population; moreover, one should strive to equalize labor conditions across the European Union. Unlike west Germans in Weststadt, Oststadt respondents rarely justified European unification by referring to the need to reassure other European countries about Germany's intentions or to a sense of obligation based on Germany's actions in the Second World War. They evaluated European integration in economic terms. This is because their historical memory is dominated by the hardships of the Communist period and by the costs of reunification rather than by events such as the Second World War. Therefore, it is not clear whether, faced with persistently high unemployment levels, east German respondents would express the same degree of commitment to European unification and the European Union as would west German respondents.

The Spanish respondents I interviewed in Quijotón and Catadell approached European unification very differently from the German respondents. Although proud of the recent economic and political modernization of Spain, they expressed the belief that Spain is well behind the rest of Western Europe in social, economic, and cultural aspects. Consequently, and in contrast to the German respondents, they tended to blame Spain and its people rather than others for such problems as unemployment rates, which are much higher than they are in Germany, east Germany included. With memories of the backwardness and ostracism that prevailed during the autarkic phase of the Francoist regime still fresh in their minds, the Spanish respondents saw membership in the European Union as the only path toward modernization and international recognition. They complained about specific policies, such as CAP, or protested against attempts to cut the allocation of structural funds to Spain, but they conveyed that they do not entertain other alternatives. European political integration will ensure that Spain belongs to a strong bloc and entails the promise of solidarity funds and an improvement in labor conditions.

Finally, the respondents in Engleton and Scotsburg revealed a major split in British society concerning Great Britain's place in the world. Some respondents proudly proclaimed that Great Britain is a rich and powerful country with a glorious history. Other respondents had a bleaker view of Great Britain. For them, the empire is a thing of the past, and what remains is a third-rate and not very rich power. When asked about European integration, most respondents, regardless of their national self-perceptions, liked or at least seemed to have come to terms with Great Britain's membership in the European Union and with the establishment of a single European market. Their views on further European political integration were not so positive, however. Neither burdened by the past nor concerned about the threat of foreign labor competition, as were the German respondents, respondents who believed that Great Britain is still a great country tended to be against further European political integration. The rest of the respondents were not enthusiastic but did not express strong qualms. Interestingly, the Euroskeptics did not rely very much on justifications that stress the economic or political advantages for Great Britain of a lack of political integration in Europe. Rather, they stressed cultural differences between Great Britain and the rest of Europe or expressed outright contempt for other European countries, especially France. Most often, however, like the majority of the British respondents, they expressed fear of losing their national identity and culture, partly as a result of not having enough clout within the European Union. These are fears that the German respondents did not seem to have. One could interpret this contrast as a sign that national self-doubt is present even among the most assertive British Euroskeptics.[7]

Public Intellectuals and European Integration

In order to understand why ordinary citizens and local elites in different countries differ in their attitudes toward European integration and European institutions, one does not need to assume that the causal factors explaining them are the same as those that explain the public intellectuals' attitudes. But public intellectuals may have played a significant role in creating these different attitudes by helping to popularize the arguments that citizens use when justifying to themselves and to others why they support or oppose particular steps toward integration (Dalton and Duval 1981).

Content analysis of 680 randomly selected lead and op-ed articles published in the quality press between 1946 and 1997 constitutes the foundation upon which my analysis of the views of public intellectuals is based.[8] I examined six daily or weekly quality newspapers, which I chose with a view to representing dominant views of public intellectuals in the three countries over time. All are influential national newspapers with a very large, if not the largest, readership in their respective countries—relative to the type of newspaper, daily or weekly. In Great Britain I selected the *Economist* and the *New Statesman*. The former is a promarket, pro-free-trade weekly, whose political views have generally been close to those of the Conservative Party. The latter, also a weekly, has closely followed the evolution of the Labour Party, from leftist to center-left, "Third Way," political views.[9] In Germany, I chose to analyze the conservative-liberal, close to the CDU, *Frankfurter Allgemeine Zeitung (FAZ)* and the liberal *Die Zeit*.[10] To analyze the views on European integration expressed in East Germany before reunification, I examined the state-controlled and national newspaper *Neues Deutschland*. In Spain, the selection was made difficult because of the dictatorship of General Franco (1939–75). For the period in which there was hardly any freedom of the press, and in order to represent conservative views in democratic Spain, I analyzed the privately owned newspaper *ABC*. For the period between 1971 and 1975, I complemented the analysis of *ABC* with the analysis of the weekly *Cambio16*, a center-left newspaper that became the most important vehicle of public dissent in the last years of the Franco Regime. After 1976, I analyzed *El País,* a center-left, pro–Socialist Party (PSOE) newspaper that revolutionized journalism in post-Franco Spain and soon became the most widely read newspaper in the country.[11]

Undoubtedly, the content of the articles in these newspapers does not fully represent the spectrum of mainstream views on European unification in the three countries. Such limitations are inevitable in a content analysis of three countries that spans a fifty-year period. While interpreting our results demands caution,[12] the generalizability of our findings in no way attenuates their significance.

■ Images of European Integration

The existence of a permanent dialogue between public intellectuals and the rest of the population with respect to European integration is reflected in the strong similarity between their images of European integration and European institutions, as well as in the meaning both groups attach to the membership of their respective countries in the European Union. The economic advantages of creating and being part of a single market, the realization that European states are too small to compete economically and militarily against the United States, the Soviet Union (Russia), and Japan, and criticism of the functioning of European institutions and of the Common Agricultural Policy (CAP) are dominant themes in the lead and op-ed articles as they were among participants in the in-depth interviews in Germany, Great Britain, and Spain. These themes are referred to in 18 percent, 19 percent, 15 percent, and 12 percent of the articles, respectively (see table 7.2).[13] One of the greatest contrasts between the themes emphasized in the lead and op-ed articles and those emphasized by participants in the in-depth interviews concerns the frequency with which the removal of border controls was mentioned. Whereas it was the fourth most mentioned topic in the in-depth interviews, it was only the thirteenth most mentioned topic in the lead and op-ed articles. The population is clearly not exclusively dependent on understandings of the European integration process provided by elites, even though the latter may contribute to focus discussions, directly or indirectly, by emphasizing certain topics.

The findings described above are consistent with the view that public intellectuals have been successful in conveying their views about European integration to the population. Indeed, the correlation between the percentages of news-

TABLE 7.2 Positive and Negative Descriptive or Evaluative Comments about European Integration in Spanish, British, and German Lead and Op-Ed Articles, 1946–97
(in percentages)

	Germany	Spain	Great Britain	Total
Understanding	3.0 (6)	1.0 (2)		1.3 (8)
Common market	18.8 (38)	14.9 (29)	19.0 (40)	**17.6 (107)**
CAP*	11.9 (24)	4.6 (9)	19.4 (41)	**12.2 (74)**
Democratic deficit*	6.4 (13)	1.5 (3)	10.0 (21)	6.1 (37)
States too small*	26.7 (54)	21.1 (41)	9.5 (20)	**18.9 (115)**
Governance	12.4 (25)	12.9 (25)	18.5 (39)	**14.7 (89)**
Against isolation*	0.5 (1)	**8.8 (17)**	7.1 (15)	5.4 (33)
Lack of voice	3.5 (7)	1.0 (2)	3.8 (8)	2.8 (17)
Modernization*	—	**5.2 (10)**	2.4 (5)	2.5 (15)
Removal of barriers	3.5 (7)	1.0 (2)	2.4 (5)	2.3 (14)
Social benefits	1.0 (2)	0.5 (1)	2.4 (5)	1.3 (8)
Sovereignty and identity*	1.5 (3)	1.5 (3)	**8.5 (18)**	4.0 (24)
Lessons of WWII*	**7.4 (15)**	—	—	2.5 (15)
Peace*	14.4 (29)	5.2 (10)	6.2 (13)	8.6 (52)
Free movt. and competition	—	—	—	—
Structural/regional funds*	1.0 (2)	5.7 (11)	3.8 (8)	3.5 (21)
N =	202	194	211	607

*Chi-square, sig. at .05 level, 2-tailed

NOTES: Understanding: Contributes to better understanding between peoples and cultures; Common market: The common market is economically beneficial; CAP: The Common Agricultural Policy is a bad policy; Democratic deficit: European institutions suffer from a democratic deficit; States too small: States are too small to face economic or military challenges; Governance: The governance of European institutions is poor; Against isolation: Membership of this country is necessary to break the country's isolation, isolation would be disadvantageous for the country; Lack of voice: The country's voice is not taken into account within European institutions; Modernization: The country will modernize as a member of European institutions; Removal of barriers: The removal of barriers to the movement of people is a good thing; Social benefits: The country's social benefits will increase as a result of membership in the European institutions; Sovereignty and identity: Membership in the European Union has or will have a negative effect on sovereignty and identity; Lessons of WWII: Membership in the European institutions will reduce misgivings toward the country; Peace: Will contribute to peace; Free movt. and competition: Free movement of workers will mean competition from foreign workers; Structural/regional funds: The structural and regional funds of the European institutions are a good thing.

paper articles and the percentages of respondents that mentioned the different themes is equal to 0.75.[14] Also, as the elite-mass diffusion model would predict, there is a greater correlation between the content of newspaper articles and the content of the in-depth interviews with local elites than between the content of newspaper articles and the content of the in-depth interviews with ordinary citizens (0.77 and 0.69, respectively).[15] It could certainly be argued that these results simply reflect the fact that both public intellectuals and the rest of the citizens are getting the same type of information from the media, which they then process with the same cultural tools. Related to this, one could say that the public intellectuals' frames are more similar to those of local elites than to those of ordinary citizens simply because of greater social proximity. Nonetheless, the fact that public intellectuals writing on European integration think more about the topic than do other citizens, even than those with the same social background, lends credence to the hypothesis that public intellectuals play some role in popularizing particular ways of conceiving of European integration. Of course, to say that public intellectuals contribute to the popularization of particular frames about European integration does not mean that public intellectuals are as effective in shaping actual attitudes toward European integration. Frames and attitudes are two different issues, and this chapter deals only with the former.

The topics that dominate the content of lead and op-ed articles have done so for a very long time. This is certainly the case for the two positive features of European integration mentioned above, which pervade discussions of European integration from the very beginning. As for criticisms of governance and CAP, they become dominant themes in the period between 1973 and 1985. In sum, we are in the presence of a fairly stable image of European integration and European institutions: the European Union as an economic and political necessity that needs reform of farming policy and style of government. This dominant image of Europe has taken longer to develop and remains more controversial in Great Britain than in Germany and Spain. One can illustrate this greater degree of controversy by examining to what extent British lead and op-ed articles see membership in the European institutions as being economically advantageous for the country. If instead of counting positive references, one focuses on negative references, it turns out that 12 percent of the British articles contain statements about the negative economic effects that membership in the European institutions would have for Great Britain. This contrasts with 2 percent in both Spain and Germany and is true for the entire period.[16] This significant finding must be taken into account when discussing the reasons for Euro-skepticism in Great Britain, although it must be noted that British respondents seemed to be as optimistic about the economic effects of membership as were the Spanish and German respondents (see above).

The images of European integration and European institutions presented in lead and op-ed articles in each country separately also mirror the images provided by respondents in each of the respective countries. In Germany, for instance, the topics of peace and consideration for other countries' misgivings about Germany play a significant role in the articles included in this sample (14% and 7% of the newspaper articles, respectively), compared to a relatively smaller role or no role at all in Great Britain and Spain (see table 7.2). In the past, German public intellectuals have been particularly sensitive to the implications of an east-west confrontation; and today, following the breakup of the Soviet Union, they are concerned with the potential for political instability in eastern Europe. In both instances, despite some debate, their conclusion is that European institutions offer the best safeguard against these potential threats. Similarly, German lead and op-ed articles have continually stressed how important it is for Germany to be an active partner in the building of Europe, in order to calm other countries' fears. In fact, the data reported in table 7.2 drastically underrepresent the degree to which German public intellectuals have been concerned about other countries' misgivings toward Germany. If instead of counting direct references to the role that German commitment to European reunification plays in other countries' fears of Germany, one counts all references to other countries' misgivings toward Germany, the overall percentage for the entire period jumps from 7.4 percent to 11.4 percent.

A comparison between the views on European integration of German public intellectuals and those of ordinary citizens reveals overall similarities but also one significant contrast, which concerns the absence of comments in the German sample of lead and op-ed articles referring to a fear of competition by lower-paid foreign workers. As I showed in a previous section, this was a very salient theme among German respondents, distinguishing them from British and Spanish respondents. It may not be a coincidental contrast. In fact, it could well reveal the care German elites take to present a good image of their country, one that is consistent with their desire to reassure Germany's neighbors. The rest of the citizens may have developed their understanding of the impact of European integration based on news that links unemployment in Germany to the greater mobility of firms and labor throughout Europe. In 1995–96, during the time I conducted my interviews, there were a good number of informative pieces in the media concerning conflict in the construction sector caused by the presence of large numbers of workers from other countries of the European Union who were working for lower wages. What we have in this case is not a process of elite-mass diffusion; rather, it is an illustration of how the European Union impacts some groups in the population, who on occasion mobilize and succeed in publicizing their grievances to the rest of the population.

The comparisons above have concerned exclusively lead and op-ed articles published in west Germany and the sample of west German respondents. These comparisons show, among other things, that the perception that Germany must contribute to efforts at European unification in order to reassure other west European countries is shared by both public intellectuals and the general population in west Germany. The description has also shown that, if anything, mentions of this topic in lead and op-ed articles increased in the period after German reunification, especially in *Die Zeit.* How is it, then, that east German respondents in Oststadt hardly ever mentioned it when discussing European integration? Is it that the east German population has not yet been exposed to the messages about European integration spread by west German public intellectuals? Is it that east German public intellectuals, despite reunification, propagate another type of messages? Or is it that the east German population is only selecting those messages that fit its current economic and political concerns and its conceptions of Germany and of German history? It is likely that the three factors combined have contributed to the contrast between west and east German respondents. First, as one would expect, the press of the German Democratic Republic, especially the newspaper *Neues Deutschland,* lost legitimation and readership upon reunification (Voltmer 1998, 90). Therefore, had post-reunification east German public intellectuals framed European integration in a different way from that of west German public intellectuals, they would have found it difficult to reach their potential public. It is telling, nonetheless, that out of seventeen lead and op-ed articles published by *Neues Deutschland* between 1990 and 1991 (the end period of my sample for this newspaper), none of them mentioned the need for Germany to reassert its commitment to European integration or advocated further European integration to reassure other countries about a too-strong Germany. Furthermore, it has been noted that the west German press, including the quality press, has encountered difficulties in setting foot in east Germany (Voltmer 1998). These difficulties, which already reveal a certain mismatch between the political preoccupations and worldviews of west Germans and east Germans, should have made it difficult for west German public intellectuals to reach the broader east German public. Finally, east German public intellectuals in the pre-reunification period presented a totally negative picture of European unification and of west Germany's role in it, a picture that has lost all legitimation but that has not yet been fully replaced by an alternative one.[17] In sum, it is quite likely that the images of

European unification currently held by the east German population have been shaped mostly by factual news information about the day-to-day functioning of the European Union and much less so by the concerns of west or east German public intellectuals. This context and the radically different set of cultural values with which the east German population was brought up would probably explain why the Oststadt respondents hardly ever expressed concern about other countries' misgivings toward a strong Germany in their discussions of European integration. More than any other contrast, the contrast between Weststadters and Oststadters reflects the role of the broader culture in shaping public discussions about European integration and the way those discussions influence the population's views of European integration.

In Spain, the desire to break the country's traditional isolation by becoming a member of the European Union and the expectation that membership in the European Union will contribute to the modernization of the country are mentioned with greater frequency than other topics (9% and 5%, respectively). A comparison of trends over time suggests that the views of ordinary citizens on European integration are shaped less by contemporaneous conditions and messages than by past conditions and messages received in the past and repeated over time. Indeed, the topics of isolation and modernization are hardly touched on by lead and op-ed pieces of the last decade, yet they were salient topics among my Spanish respondents. One needs to go as far back as the 1959–73 long decade to find the largest concentration of references to these two thematic arguments. Nineteen percent and 16 percent, respectively, of the lead and op-ed pieces in that period contain references to the topics of isolation and modernization.[18] The frequency with which these themes recur in Spanish lead and op-ed articles is somewhat higher than it is in Great Britain, and it becomes more so if one adjusts for the difference in the number of topics mentioned in Spanish and British newspaper articles (10% and 6% of all mentions touch on the topics of isolation and modernization in Spain, compared to 6% and 2%, respectively, in Great Britain). Needless to say, these topics are all but absent in discussions of European integration in German lead and op-ed articles. A detailed reading of the newspaper articles would reveal that in fact, this concern about isolation and modernization is part of a deeper concern about increasing Spain's role and prestige in the world scene that clearly distinguishes Spain from Great Britain and Germany.

Upon comparing the content of Spanish lead and op-ed articles with that of the in-depth interviews conducted in Quijotón and Catadell, one finds that the CAP and the structural and regional funds are much less salient in the former than in the latter. This contrast is less telling for the structural and regional funds, however, than for the CAP. Despite their low salience as a topic of discussion in lead and op-ed articles for the entire period, the structural and regional funds were the second most mentioned theme in articles published in the 1986–97 period in *ABC* and *El País*. The CAP is another matter: the percentage of lead and op-ed articles in the Spanish sample devoted to this topic has been very small, even in the 1986–97 decade. As I said for the topic of foreign worker competition in Germany, this is an instance of abundant news that has not been echoed by public intellectuals. Because of the significant role of agriculture in the Spanish economy and the effective mobilization of Spanish farmers to defend their interests against the European Union, the CAP is very often in the news, and it was so in the period 1996–97, when I conducted the in-depth interviews in Spain. The contrast between the low level of debate about the CAP among public intellectuals and the high salience of this topic among the general population reveals different sensitivities as to what is important about the European Union. The fact that the CAP is less mentioned in the Spanish lead and op-ed articles than in the British and the German is also interesting, but that can be easily explained by the fact that Spain was not yet a member of the EEC in the late 1970s, when most disputes concerning the reform of the Common Agricultural Policy took place.

Several features distinguish British leads and op-eds from those published in German and Spanish newspapers. First of all, less emphasis is placed on the belief that European institutions are needed for European states to be more competitive and secure, both economically and militarily. Only 10 percent of the articles mention this topic, compared to 21 percent and 27 percent in Spain and Germany, respectively. A look at trends over time suggests that this has always been so, a fact that contrasts with the large number of British respondents who mentioned this theme during the in-depth interviews. These findings ought to be examined jointly with the previously noted tendency of British lead and op-ed articles to attribute negative economic effects to Great Britain's membership in the European institutions. Overall, the British intelligentsia does not appear to have been persuaded as much as the Spanish or German intelligentsia of the economic and political advantages of a supranational organization such as the European Union. Another related feature that distinguishes British from German and Spanish articles is the relatively high frequency with which the British ones refer to the negative consequences that European integration and membership in the European institutions have on national sovereignty and national identity. German lead and op-ed articles also frequently discuss the topic of sovereignty and identity, but only to point out that a transfer of sovereignty is a good thing. Indeed, a remarkable 22 percent of German articles mention the transfer of sovereignty as a good thing (compared to 12% in Great Britain and 4% in Spain), whereas only 2 percent mention it as a bad thing (compared to about 9% in Great Britain and 0% in Spain).[19]

Sovereignty is therefore a highly debated topic among British and German public intellectuals writing about European integration and an irrelevant issue in Spain. But whereas the transfer of sovereignty has been highly contested in Great Britain, it is overwhelmingly supported in Germany. The examination of temporal trends shows that the period of greatest opposition to a transfer of sovereignty in Great Britain was the period between the creation of the EEC and EURATOM (1957) and Great Britain's accession (1973). Between the British accession and the signing of the Single European Act (SEA) (1986), the debate continued, but comments favorable to a transfer of sovereignty became more numerous than those against it. Finally, in the last decade, the intensity of the debate has declined considerably, and comments in favor of a transfer of sovereignty have more than tripled over those against a transfer.[20] This is yet another instance in which popular attitudes toward European integration do not exactly correspond to what is said by public intellectuals at that particular moment: while British respondents often argued against European integration by criticizing the erosion of sovereignty and, more significantly, of their identity, lead and op-ed articles in the center-right and center-left quality press had basically abandoned this type of discourse. The debate on the European Monetary Union is a case in point. Both writers for the *Economist* and the *New Statesman* have expressed lukewarm support for the single currency, but criticism of the process of monetary unification has not been made on grounds that it violates British sovereignty. Rather, it has been directed at the convergence criteria and the lack of popular legitimacy. When British respondents criticized the EMU, however, they mostly referred to issues of sovereignty and identity.

In sum, lead and op-ed articles about European integration in Germany, Great Britain, and Spain commend its economic and politico-military aspects and criticize its governance aspects and CAP. These evaluations differ, however, from country to country. Meanwhile, the comparison between the results reported in this section and those corresponding to the analysis of the in-depth interviews with ordinary citizens and local elites reveal great similarities. These similarities reflect great stability in the frames that have distinguished Great Britain, Germany, and Spain with reference to European integration since the late 1940s and are consistent with the assumption that public intellectuals contribute somewhat to shape the population's way of conceiving of European integration. The fact that ordinary citizens often express topics that are unlikely to be discussed in merely informational pieces, such as the need to calm other countries' fears in the case of west Germany, indicates that the impact on public opinion of public intellectuals' views is not limited to the most politically aware segments of the population.

■ Unfinished Agenda

Because of the consistency of the findings above with elite-mass diffusion theories, one would be tempted to stop the analysis here and conclude that international differences in support for European integration simply reflect differences in the attitudes of these countries' elites. These differences between elites could in turn be explained by the political and economic constraints faced by Great Britain, Germany, and Spain during the second half of the twentieth century. This apparently compelling account would still be incomplete, however, because it does not completely explain, for instance, why sovereignty, isolation, modernization, and concern for other countries' misgivings about Germany have become so central in the public intellectuals' approach toward European integration in Great Britain, Spain, and Germany, respectively. Second, it is far from clear that the distinguishing arguments provided by public intellectuals would have struck a chord in the collective consciousness had these arguments failed to connect with broader concerns in their societies or tap into deeply felt national self-images. An explanation of international variation in support for European integration focused on the role of elites would thus be incomplete. It needs to be complemented by an argument about the role of broader cultural preoccupations in shaping people's ways of seeing European integration and in facilitating the public intellectuals' role in popularizing these frames. This issue will be explored elsewhere.

NOTES

I would like to thank Mabel Berezin, Berit Dencker, my colleagues at the Department of Sociology of UCSD, and researchers at the Wissenschaftszentrum Berlin für Sozialforschung for comments on parts of this chapter. I would also like to express my gratitude to the Alexander von Humboldt Stiftung, the Program for Cultural Cooperation between Spain's Ministry of Education, Culture, and Sports and United States Universities, the Center for German and European Studies at the University of California, Berkeley, and the Academic Senate of UCSD, for funding provided for this research project.

1. Recent work in the social movements literature emphasizes the need for collective action organizations to take people's values and beliefs into account. The concepts of frame alignment and frame resonance are particularly important in this respect (see, e.g., Snow et al. 1986; Snow and Benford 1992; Babb 1996).

2. Wessels (1995), using a sophisticated methodology, demonstrates convincingly that party elites and party supporters influence each other and that the influence of the former is greater than that of the latter.

3. The cities' names have been changed to maintain the anonymity of the respondents. The actual names can be obtained from the author.

4. Constrained as I was to working with small samples, because of the ethnographic character of my project, I consciously sacrificed some intracity representativeness in order to ensure that compositional factors such as age and education did not affect the intercity and international comparisons. This sample selection procedure also made intracity comparisons between respondents in different age and education groups possible. Note 7 below shows, however, that there are good reasons to view the results that follow as a valid reflection of the main contrasts between the British, German, and Spanish frames about European integration.

5. A full comparative analysis of this and other sources appears in Díez-Medrano 2003.

6. Tables available on request, which may be sent to jdiezmed@weber.ucsd.edu.

7. Although these results have been obtained from small samples that were not fully representative of the British, German, and Spanish populations, they can be taken as an adequate reflection of the main contrasts between the Spanish, British, and German frames about European integration. The number of interviews that I conducted in each of the cities—between 24 and 29—allowed me to attain closure in terms of the list of arguments respondents use to justify their views of European integration and the European Union. It is highly unlikely that topics that were not mentioned or were seldom mentioned in a particular city would have been mentioned or mentioned much more often with larger or more representative city samples. Thus, while it would certainly be risky to make precise population inferences about the relative salience of the various themes discussed here, it is safe to make inferences about the themes that distinguish British, German, and Spanish frames about European in-

tegration from each other. One can confidently rule out, for instance, that arguments about the lessons drawn from the Second World War would have been expressed in Great Britain or Spain with a larger or a different type of sample in these two countries. Meanwhile, the theme of national identity was clearly concentrated in the United Kingdom, whereas the themes of modernization and isolation were clearly concentrated in Spain. The fact that these distinguishing British and Spanish frames were formulated with similar frequencies in the two cities of each of these countries respectively strengthens the conclusion that they are indeed frames that serve to distinguish British, German, and Spanish views of European integration.

Alternative sources of information further buttress the conclusion that the Second World War, national identity, modernization, and isolation are indeed frames that distinguish German, British, and Spanish frames about European integration and that can thus be used to account for contrasts in support of European integration between these three countries. The salience of modernization and isolation in the Spanish population's frames, for instance, is confirmed through data from a national representative survey conducted in March 1999 by ASEP, a prestigious survey research institute. Respondents were given open-ended questions in which they were asked to justify their attitudes toward European integration and Spain's membership in the European Union and to list the aspects they most liked and disliked about the European Union. The rank-order correlation between the frequencies with which the sixteen different themes in table 7.1 were mentioned in this national representative survey and those corresponding to my city samples is 0.6. Furthermore, people's expectations that European integration and the European Union will contribute to Spain's modernization was the sixth most mentioned topic (21.2% of the respondents mentioned it), and the desire to break with Spain's traditional isolation was the ninth most mentioned topic (5.1% of the respondents mentioned it). These results mimic those obtained in my Spanish city samples, for modernization and isolation were the seventh and ninth most mentioned themes, respectively. Also, the fear of losing the country's national identity and sovereignty was hardly mentioned by participants in the Spanish survey. Only 0.7 percent of the sample mentioned it, which confirms that fear of losing the national identity, a major component of the British frame about European integration, is not a salient feature of the Spanish one.

The fear of losing the national identity because of membership in the European Union is indeed a unique and distinguishing feature of the British approach to European integration. The 1999 Eurobarometer 51.0 asked respondents in a direct, closed-ended question whether they were afraid of losing their national identity and culture as a consequence of the European Union. It should come as no surprise that the country percentages of respondents who expressed this fear are higher than they are in my city samples or in the 1999 Spanish survey, in which they were obtained through open-ended questions that were asked as follow-up to general questions about European integration. Nonetheless, the survey results show that whereas 64 percent of the British respondents expressed such fears, only 44 percent, 42 percent, and 35 percent in west Germany, east Germany, and Spain, respectively, did so.

8. I collected all lead and op-ed articles on European integration in the selected newspapers for all or selected years between 1946 and 1997. In the United Kingdom, I searched all issues for every year from 1946 to the end of 1997. Similarly, in Spain I searched all newspaper issues for the newspapers *ABC, El País,* and *Cambio16* for the same period. In Germany, I searched every even year for the *Frankfurter Allgemeine Zeitung* and every odd year for *Die Zeit,* also for the same periods. From this body of articles I drew a systematic and stratified sample for each country whose content I analyzed. I thank Silvio Waisbord, Natasha Unger, and Paloma Díez Hernando for their assistance in collecting this information.

9. For descriptions and analyses of the British press during this fifty-year period, see Seymour-Ure 1968, 1991; Tunstall 1996; Anderson and Weymouth 1999.

10. For descriptions of the histories, market presence, and content of these newspapers, see Kiefer 1992.

11. For the recent history of the Spanish press, see Giner 1983 and Alférez 1986.

12. Concerned about the degree to which some of my selected newspapers provided an adequate view of dominant center-right and center-left frames about European integration, I analyzed editorials and op-ed articles from the *Times* (Conservative), *Frankfurter Rundschau* (Social-Democrat), and *Arriba* (Falangist), for the entire period examined here (except for *Arriba,* for which my sample was limited to the 1950s). My results did not contradict the findings presented in this chapter with respect to frames about European integration that are present in the *Economist, Die Zeit,* and *ABC,* respectively.

13. One must be careful with the comparisons because of the different conditions under which information was obtained. The interviews were directed, to the extent that the same questions, covering a wide variety of topics, were asked of every respondent. The articles' content is decided by their authors. Therefore, one should not expect very high percentages for these categories.

14. Significant at the .05 level. $N = 16$.

15. Both are significant at the .05 level. $N = 16$. One observes the same contrasts by comparing the correlation between the content of newspaper articles and the content for informed respondents with that between the content of newspaper articles and the content for uninformed respondents (the correlations are 0.71 and 0.66, respectively).

16. Table available on request (see note 6 above).

17. The articles during the period emphasized four negative consequences of European integration: transfer of sovereignty to European institutions, rising prices, declining social standards and worsening conditions for the workers, and the impossibility of achieving German reunification. A frequent variation of this last theme consisted of predicting that the Federal Republic would take over the German Democratic Republic. Moreover, the paper did not miss a single occasion to point out the imperfections of the EEC, be it national disputes over the CAP or over the budget, the monetary crises of the 1970s, the expected decline in the living standards of workers, or the resistance of the working classes in Great Britain

and Scandinavian countries to membership in the EEC (Díez-Medrano, forthcoming).

18. Tables available on request (see note 6 above).
19. Tables available on request (see note 6 above).
20. Tables available on request (see note 6 above).

EIGHT

Europe and the Topography of Migrant Youth Culture in Berlin

LEVENT SOYSAL

"Wer nicht zu Europa gehört"

In its 11 October 2000 issue, the intellectual opinion weekly *Die Zeit* published excerpts from Helmut Schmidt's new book on the future of the European Union.[1] The piece, entitled "Who Doesn't Belong to Europe," a variation on the ever popular political and intellectual exercise "who-belongs-to-Europe," focuses on four countries: Russia, Ukraine, Byelorussia, and Turkey, the first three bordering the current eastern end of what was formerly called Western Europe, and the fourth unmistakably located on its Middle Eastern margin. In this article, the elder statesman Helmut Schmidt takes it upon himself to provide the answer to the laborious question and resolve the conundrum of European expansion.

The task, of course, is anything but easy. Schmidt begins his excursion by reminding his readers of a dispute he had with Charles de Gaulle, who spoke of a "Europe from the Atlantic to the Urals" and the "Europe of Fatherlands." Schmidt dismisses the first of de Gaulle's visions as a mere colloquialism of the "school-book geography" kind (although on most maps of Europe the Ural Mountains still mark the territorial end of Europe), and for the latter he expresses his affinity, both then and now. He then asks rhetorically whether there is not a "European Culture" that can be defined side by side with "national identities" and "incorporated into [people's] consciousness." It is from this premise of larger European culture that Schmidt begins. He moves to determine who belongs to Europe, or rather he attempts to draw cultural borders to exclude Russia, Ukraine, Byelorussia, and Turkey from the realm of Europe. His definition of culture is quite prosaic: the usual collection of religion, science, literature, music, architecture, and so on, and political ideals of individual freedoms and rights. (Although Schmidt includes in his list "the principle of equity embedded in welfare state," this does not seem to be a shared European ideal at the moment, given that European states are distancing themselves from the notion of welfare under the spell of market economy and given Schmidt's own emphasis on privatization as a condition of Europeanization later in the article.) As expected, for Schmidt the list of positive aspects of "the European culture" outweighs its negative aspects ("for example, the inquisition, witch hunts, torture, antisemitic pogroms, and other Europe-wide atrocities"), and he ventures to choose those who belong and those who do not from the list of nation-states waiting at the gates of the European Union.

Without feeling a need for much explanation, Schmidt effortlessly includes in the European culture realm *(Kulturkreis)* Poland; Hungary; the Czech Republic, and "likewise" Slovenia; the Baltic republics Estonia, Lithuania, and Latvia; and the Balkan states Romania, Bulgaria, Slovakia, Malta, and Cyprus, "except its Turkish section." Because of their "cultural character," they either belong to Europe or "should be integrated into the "cultural continuum of Europe." The ex-Soviet republics Russia, Ukraine, and Byelorussia fall outside the European culture realm because of their history of almost a hundred years under totalitarian regimes (and to an extent their religious differences as Orthodox Christians), although Schmidt takes pains in acknowledging the contributions of Mussorgsky, Tchaikovsky, Stravinsky, Prokofiev, Shostakovich, Pushkin, Gogol, Dostoevsky, Tolstoy, and others to the making of European culture. Turkey falls off the map because of its centuries-long entrenchment in another cultural realm, that of Islam, and its shortcomings in the realm of human rights. He agonizes over his personal relationship to Turkey ("I visited Turkey many times, also as a private person") and the geopolitical concerns of Europe (Turkey's membership in NATO, its role concerning the stability of the Middle East and the Caucasus, and its trade relations with Europe). Nonetheless, he places Turkey on the other side of the border, a "friendly neighbor" with a different culture. In the case of Russia, Ukraine, and Byelorussia, he allows for a possibility of reintegration, albeit after a generation.

(Re)Mapping Europe

My interest in Helmut Schmidt's ruminations does not stem from a concern to exploit them to point out the futility of what passes as intellectual deliberation on culture, history, and geopolitics in public debates over Europeanization and Europeanness. Neither it is my intention to refute Schmidt's assertions by providing empirical evidence to the contrary. This is all too obvious and all too facile. In fact, Schmidt himself is aware of the facts of history that complicate his story and render his conclusions untenable. Yes, for instance, he accepts that Islam has had corporeal and cultural presence in Europe both then (i.e., "Cordoba was Muslim until Spanish Reconquest") and now (i.e., "Around us live over three hundred million people of Islamic culture and religion" and "in Germany and France live three million Muslims, and another one and a half million in England, and a half million in the Netherlands"). What is intriguing for me in Schmidt's exercise in border management is one invective he poses in passing: "Those, who wish to admit Turkey into the European Union regardless, must know with which arguments they will later reject the possible application of Egyptians, Moroccans, Algerians, and Libyans."

In this seemingly defiant and rational enunciation, I assert, lies the predicament of defining Europe. The Ural Mountains may appear as a feasible and tangible place to mark the end of Europe and the beginning of Asia (after a generational changing of the guards in Russia), but once Turkey is in, as Schmidt professes, no line will be left to demarcate Europe. In fact, it could be safely argued that even without the formal membership of Turkey in the EU, the lines that divide Europe from the rest are already blurred. With movements of peoples, goods, and cultures, the southern shores of Mediterranean Europe extend to the northern shores of Africa and the western shores of the Middle East. Within numerous institutional arrangements (e.g., the European Union, the European Commission, the Council of Europe, the European Court, the North Atlantic Treaty Organization, the European Free Trade Association, the Congress of Local and Regional Authorities of Europe, and the Council of European Municipalities and Regions), the states and cities within and outside the European Union are entangled in discernible and substantive ways. As signatories of agreements (i.e., the European Convention on the Protection of Human Rights and

Fundamental Freedoms, the European Social Charter, and the Customs Union Agreement), nation-states commit themselves to follow the dicta chartered as European. At annual rituals of culture and sports (i.e., the Eurovision Song Contest and the European Football Championships), nations and cities perform and compete on the stages of Europe. In this sense, Europe is not one but many, depending on the countries and regions included in the permutations of arrangements, treaties, and connections. And in varying degrees, Turkey, as much as Germany, France, and Greece, and as well as Israel, Morocco, and Azerbaijan, belongs to this multifaceted Europe as a member of institutional structures, as a signatory of treaties, and as a participant in the making of the European.

Accordingly, it is only appropriate to characterize Europe as a supranational entity, taking shape in the discontinuous geography of Europa and shaped by enfolding inconsonant politics and economies of regional, national, and global order. This Europe does not denote a readily identifiable cultural or geographical area but a social, institutional, and cultural formation in the making. It includes but is not exclusive to the European Union, and as a collection and network of nation-states, regional entities, and cities, it is conclusively connected to its Mediterranean, Balkan, Middle Eastern, and eastern European margins. This Europe is also home to foreigners belonging to a wide array of differentially organized membership categories, including third-country (non-EU) citizens, asylum seekers, dual citizens, holders of various temporary and permanent residency rights, and illegal aliens. The foreigners of Europe are subjects in a complex terrain of exclusions and inclusions, contention and accommodation, and disenfranchisement and membership. As Europe becomes a supranational entity, larger, world-level tendencies toward the "'denationalization of the state,' [and] 'privatization of nationality'" (Bauman 1990, 167; see also Sassen 1996; Soysal Nuhoğlu 1994) compound the emerging notions of Europeanness and the proliferation of regional and minority identities. While Europeanness presents encompassing, inclusive formations, regional and minority identities project cultural pluralisms, as well as cleavages, within the confines of the newly imagined Europe.[2]

This working definition of Europe, which frames my arguments in this chapter, is necessarily ambiguous. As Castells suggests, though there is an undeniable movement toward the unification of Europe, "the content of this unification, and the actors involved in it, are still unclear, and will be so for some time" (1998, 331).

Castells, of course, is neither alone nor peculiar in his assertions. Soysal Nuhoğlu, drawing attention to the contentious nature of the processes of Europeanization, similarly contends that "the script of Europe is still open to modifications and re-writing, and it may never end with a coherent narrative" (2002, 281). And in Varenne's rather celebratory words, "we have here a large scale experiment in culturing (cultivation?) of history," presenting "a singular problem for the collective imagination of humanity" (1993, 236–37). What is impressive is that in all three accounts, the emphasis is on the process, with its uncertain outcomes, and the (in)formal sites of debating and making Europe. The process is better characterized "as a debate rather than as a blueprint," remarks Castells (1998, 331); Europe expressly "happens" through "non-state associational and networking practices," maintains Soysal Nuhoğlu (2002, 273); and, for Varenne, intellectuals, who give voice to affirmative or skeptical "grumblings" in matters of Europeanization, are implicated in the act of "building, constructing, perhaps, even creating and inventing" Europe (1993, 237). Last but not least, we should not forget the opposition to the idea and process of Europeanization, within and outside the EU.

It is in this spirit that I approach the project of the remapping of Europe that this book attempts to chart and add another voice to the argumentation regarding the prospects and processes of Europe-making. This chapter aims to remap the topography of migrant youth culture in Berlin and examines the ways in which the cultural projects of Turkish migrant youths challenge the established categories of territorialized identities and concurrently participate in shaping social and cultural matrices of belonging in Europe. As a much revered convention, the territorial division of culture assumes one-to-one correspondence between place and identity (of the ethnic/national kind, as in Germanness, Turkishness, Europeanness)—hence the incessant debate over the borders and border controls. In this model, Turkish youths ***naturally*** occupy Turkishness, and places of their livelihood in Berlin appear as ***actual*** sites of dislocation. At best, they are located in liminality—a place of neither here nor there, being "betwixt and between," and having "few or none of the attributes of [its] past or coming state" (Turner 1974, 232). In the territorially delimited world of migrations (and immigration studies), liminality only amounts to cultural displacement and perpetual identity crisis.[3] Their cultural place in the Turkishness of tradition presumed inherent but no longer workable because they are migrants in a foreign land, and their passage to the Germanness of modernity considered virtually handicapped because of the territorial closure of cultures, the Turkish youths remain in situ in uncertainty—in the ***un***becoming spaces of categorical in-betweenness.

Absent from the analyses are places where they have, make, and debate culture: neighborhoods (Kreuzberg), cities (Berlin), and states—of the national kind (Turkey, Germany) and the supranational kind (Europe). These are not simply geographical locations but, in the words of Certeau, are "spatialized" ("practiced") places, "actuated by the ensemble of movements deployed within [them]," and "function in a polyvalent unity of conflictual programs and contractual proximities" (1984, 117). They are spatialized through enactments of eclectic, discrepant cultural and societal projects, designed and envisioned by protagonists, ranging from citizens/cityzens and collectivities to state actors, some more powerful than others. Migrant youth culture happens in and at the intersection of these practiced places, informed by the contemporary discourses and signs of identity and culture and specificities of the places located in the maps and history of Europe.

In this chapter I pursue two related goals. First, I seek to elaborate an argument against in-betweenness and locate migrant youth culture in Kreuzberg, in Berlin, and in Europe. There, located in Kreuzberg and engaged in cultural projects spanning Europe, Turkish youths address their diverse audiences, articulate their political orientations, and engender spaces of self and belonging. Their cultural projects are not revivals of an essential Turkishness (or Islam) in response to alien formations of modernity. Rather, I argue, their projects contribute to the remapping, remaking of the new Europe, unsettling the conventional configurations and conceptions of belonging and otherness. Second, I seek to explicate the trajectories of Kreuzberg, Berlin, within multireferential projects of unification of Berlin, Germany, and Europe in the aftermath of the fall of the Wall. Not only are these intertwined projects attempts to supplant old boundaries and define new places in the extant maps of the globe. They also entail visions of the future for cities and states, emancipated from uninvited pasts of nations (wars and destruction) and unburdened from inadmissible presents (inequalities and exclusions) in a world increasingly imagined as global.

Pursuing this two-pronged goal is a means to an end: for migrant youth culture, its prospects and limits, becomes visible and meaningful when situated within the larger projects of (re)mapping. Otherwise, we continue to reify in our narratives timeless and placeless identities, stranded in perpetuity in the nowhere of in-betweenness, and we lose sight of their competencies and investments. The social and cultural economy of migrant youth culture belongs to the same epoch as — in Johannes Fabian's words (1983), it is "co-eval" with — the correlative praxis of envisioning the new Berlin and the new Europe.

I open my narrative of migrant youth culture by locating Turkish migrant youths in Berlin. I identify the institutional settings and public spaces in Berlin within which youth culture is produced and performed. Then I turn my analytical focus to the cultural stages of NaunynRitze, a youth center in Kreuzberg, and document the contemporary repertoire of migrant youth culture. The inventory of youth culture registered therein will reveal the connections of youth culture to the making of Europe both as a transnational entity and as an identity.

■ Berlin and Its Migrant Youth

Since the unification of two Germanies a decade ago, the city of Berlin has come to occupy an increasingly glamorized place in the fin-de-siècle imaginary of Europe, as well as of the world. As projected in official and popular visions, Berlin is a (trans)national actor in its own right — tropically, but nevertheless an actor. It is named "the Open City," with a proper definition: "'Open' means: 'ready' for change, receptive, forward-looking, open to what is strange, different, new. Open doors and buildings — so to speak a metaphor for the processes of transformation intended to create a redesigned European metropolis of mediation, communication and exchange" (*Berlin* 1999, 6). This stately vision of "Open City Berlin," prescribed by the Senate of Berlin in association with various nongovernmental agencies in a recently published glossy guidebook to "the City on Exhibition," endorses a cosmopolitan future for the city and offers that future as a cultural and social project — a "New Berlin" for the new millennium. The future of the "New Berlin" is clearly located in the "New Europe" and anticipates a city of diversity (a "variety of architectural styles, squares and city districts") and diversity of "people who live in it and with it" (6).[4] As is the case with any actor, Open City Berlin has many equally pronounced identities: *Hauptstadt* of the unified Germany, *Kulturstadt* in a unified Europe, and *Weltstadt* in a cosmopolitan world. And, in its many guises, Open City Berlin underwrites utopias and dystopias, past and present, invoking a thematic spectrum of diversity, pluralism, tolerance, and racism.

Berlin is also the symbolic and physical site of German unification—and a proxy for unifications of East and West the world over. As the unification of two Germanies has progressed, the public debates have necessarily involved and accentuated questions of cultural conflict and identity—both between Germans and others (migrants, refugees, and Gypsies), and between Germans themselves (Ossies and Wessies). Thus, in the context of unified Berlin, the struggle over identity readily goes beyond the simple dichotomy of *Us* (Germans) and *Them* (migrants = Turks). Politics of and claims to identity blur the demarcations of Germanness, fracture national boundaries, and authorize new local and transnational imaginations: inclusive or exclusive Europeanness, pluralisms of various sorts, and new migrant identities.

This Berlin, the city, in its capacity as the locale and the actor, simultaneously delimits and conditions the lives of Turkish migrant youths. The current population of Berlin is about 3.3 million, of which approximately 13 percent is foreign, with migrants from Turkey comprising approximately 4 percent of the total population.[5] Of the Turks in Berlin, 36 percent are between the ages of ten and twenty. Turkish youths are ***cityzens*** of this metropolitan city, with varying degrees of rights ranging from citizenship to undocumented residency, and in contact with peoples and cultural flows from around the world. They live in Berlin, a place that is distinct from Germany. This distinction is a legacy of divided Germany as much as it is an extension of Berlin's projected cosmopolitanism.

In this Berlin, Turkish youths occupy Kreuzberg, a district known for its eccentricity, alternative lifestyles, and real action. While Turks live and work in other parts of Berlin, Kreuzberg has a special place in the imaginary of Berlin as a metropolis. Kreuzberg, and its main crossroad, Oranienstrasse, appear in the portrayal and self-presentation of the city as the locus of hip and diversity. This vision of Kreuzberg is extremely popular and quasi-official. Not surprisingly, the municipality of Kreuzberg aptly refers to the district as Kiez International, a close-knit international neighborhood of sorts. As prescribed, Kreuzberg is said to belong to the "marginals," along with punks and artists, for it epitomizes the exotic, the ghetto, and the hip. Living at the heart of Berlin, where the local, national, and global intersect and coalesce, Kreuzberg's youths respond to the challenges and discourses offered by the urban cosmopolis, ranging from conformity to social protest. Berlin's (and Kreuzberg's) urban landscape accommodates, and colors, the everyday experiences and life-cycle projections of the Turkish migrant youth.[6]

What I have been trying to do is not to establish (or glorify) the diversity or cosmopolitanism of Kreuzberg or Berlin as such. Neither diversity nor cosmopolitanism simply comprises empirical facts on the ground, so to speak. On the contrary, they are societal projects, which are authorized by legitimating discourses and economies of our times and enacted by a multiplicity of state and non-state actors. Accordingly, through the projects of diversity and cosmopolitanism, Berlin and Kreuzberg become "spatialized (practiced)" locales (Certeau 1984) and frame the formations of youth culture, which in turn contribute to the constitution of the very diversity and cosmopolitanism anticipated and articulated by the project of Open City Berlin. In other words, the enactments of migrant youth identities and subcultures are shaped by the resources and the climate afforded by the habitus of culture in Berlin, while at the same time their performances of diversity weave the very fabric of the city's cultural "habitus" (Bourdieu 1977) — as projected in the definition of Open City Berlin.

Much of the literature overlooks the promises and adversities offered by the city, despite their significance in shaping migrant youth culture, and simply concerns itself with timeless (and necessarily Turkish) traditions and cultural formations. Without allowing for historical predicaments, the terrain of Germany becomes congruent with Germanness (the land of natives), and the contemporary geography of Turkey embodies Turkishness (the place the foreigners come from). The migrant's roots and home immanently are elsewhere. Her presence, praxis, and culture effectively remain temporary and restless against the inherent permanence of the native existence. In this seemingly intuitive scheme, migrant youths, the unequivocal second generation of migration studies and policy papers, are deemed to be ***in-between,*** or locked in a state of neither-here-or-there. They invariably appear as stranded and torn between ***tradition,*** that which is of home, parents, and distant places and pasts, and ***modernity,*** that which is of streets, social workers, and here and now.

Being evicted from everyday life and placed in the nowhere of in-betweenness is a punishing occupation. In endless tales of integration, scholarly attempts are made to define migrants' social and cultural situation by way of furnishing empirical content to the otherwise inventive category — the second generation. The science of psychology probes the degree of their adjustment and the ways they cope — rather, the ways they do not cope — with the changes that they encounter in their new culture. Other social sciences set out to measure their integration in units of educational attainment, unemployment, language competency, religious activity — as if these variables self-evidently reflected cultural conflict or disengagement from social life. In more sinister versions of statistical exercise, the migrant youths appear as relentless agents of radical or criminal otherness.

For this, I argue, we should "write against" (Abu-Lughod 1991) in-betweenness. Firmly anchored in daily rhythms of school, work, and street, the condition of migrant youths defies assertions of in-betweenness. They are not located in the shadows of a precarious nowhere, as the model dictates. On the contrary, they inhabit the familiar of here-and-now in Berlin, Germany, Turkey, Europe, and the transnational spaces of youth culture. They confidently conduct their daily life in the social spaces of Berlin, negotiate tensions and anticipations inscribed in life course narratives, and engage in the civic and cultural projects of their times.[7]

Particularly significant for their cultural projects are the organizational resources available to them as youths — cultural and recreational centers, clubs, cafés, sports associations, and numerous social and educational programs. These are well attended and, despite recent budget cuts, most receive ample support from the Berlin Senate, financially and otherwise. In fact, Berlin is the Holy City of youth organizations. For instance, the Foreigners Office of the Berlin Senate lists in one of its publications about 180 "inter-cultural" associations in East and West Berlin (*Ausländerbeauftragte* Berlin 1992). Of these, 45 are specifically youth organizations. *BERTA,* a widely distributed handbook designed for young women, registers the addresses of more than 400 youth clubs scattered throughout Berlin, about half of which provide special services for girls (*Berliner Taschenwegweiserin* 1995). In an informational booklet designed to inform immigrants and refugees, the municipal government of Kreuzberg catalogs 160 organizations in its district, ranging from counseling bureaus, cultural centers, and adult education and job training schools to kindergartens and women's coffee houses (Kiez International 1991). And in a booklet on *Jugendarbeit* (youth work), Kreuzberg's government proudly mentions 12 youth centers it operates, along with a street-work outfit called *Koko,* a job-exchange center, and a mobile team for condom distribution (*Bezirksamt Kreuzberg* 1990). These youth centers include facilities such as music and dance studios, painting, graffiti, pottery workshops, exercise centers, and conference rooms, as well as recreational spaces with billiard tables, boom boxes, TV sets, a bar serving tea, coffee, and soft drinks, and a small kitchen where the youths learn to cook and clean up after themselves. In districts with large immigrant populations such as Kreuzberg, Turkish youth comprise the majority of the membership of the centers. According to a 1997 survey conducted by Berlin's Foreigners Office, on the average every fourth Turkish youth in Berlin is a member of an association or club. While the membership figures for male youths are quite high (about 38%), they are lower for females (only 8%) (Pressemitteilung 1997).

In what follows I turn my attention to a youth center in Kreuzberg, NaunynRitze, and provide a brief ethnographic rendition of two events from the world of migrant youth culture in Berlin—a hip-hop festival and a multinational antiracist music project, both of which were performed on public stages, exploiting the institutional resources I have mentioned. These events imply different images from the instinctive sketches of a lost and disoriented second generation. As will be evident, Turkish youths do not form isolated, marginalized islands of subculture in a foreign land. I argue that in exercising their membership and otherness in public spheres, they partake in the construction of the social and cultural landscape of the new Europe and enter into dialogues with the global flows of youth culture. The events I narrate do not exhaust the range and content of migrant youth activity in Berlin. However, they evidence trends and orientations that distinguish the contemporary topography of migrant youth culture, the design and emblems of which resonate with the ethos of diversity, tolerance, and identity embodied in the projects of New Berlin and New Europe.[8]

■ Performing Hip-Hop and Diversity in Kreuzberg

In March 1994 Kreuzberg was a stage for a youth festival called Street '94. The festival was organized under the auspices of NaunynRitze, located at the center of Kreuzberg, just off the subway station Kotbusser Tor. As a self-acclaimed home for cosmopolitan hip, NaunynRitze is Kreuzberg's showcase for multicultural youth performances. Street '94 was cosponsored by Kreuzberg's municipal government and Berlin's Ministry of Youth and Family, and the financial support for the project came from state and municipal sources, as well as Berlin's biweekly program magazine *Zitty,* the highly popular radio station *Fritz,* and private corporations. As it was conceived, Street '94 was a special project in the program of another Berlin-wide youth project staged at the time—an upscale Youth

Art+Culture project called X-'94, with the subtitle "50 Days to Blow Your Mind." X-'94 was a production of (former East) Berlin's Akademie der Künste (the Academy of the Arts) and, in a way, signaled the academy's entry into the world of hip. In the grand design of X-'94, Street '94 represented the cool art of the street, subcultural undercurrents of the metropolis, and the raw skill of ghetto boys and girls.

And Street '94 was true to its projected image. It was located in Kreuzberg, the ghetto, with a touch of hip and avant-garde. Publicity posters and graffiti in Kreuzberg radiantly displayed the mottos of the festival: "To Stay Here Is My Right" and "We All Are One." In conformity with the premise of hip and hip-hop, the name and mottos of the project prominently invoked English. The festival, two months of extensive activity, revolved around a street art exhibition, workshops on graffiti writing and rap music, dance parties with rap and ethno-pop bands, and open-air screening of films such as *Boyz 'N the Hood* and *Menace II Society.* The day-to-day activities of the festival were carried out by a group of youths from NaunynRitze. Neco, a Turkish youth, and Gio, an Italian, were responsible for running the show. They were both self-made street artists, who expanded their repertoire as painters. It was Neco's dream to make a film on the migrant youth experience in Germany—a film just like the award-winning French ghetto story *Hate.* Gio had his works exhibited in a major Berlin gallery as part of X-'94. In a press release, Neco and Gio explained the vision of their posse, "To Stay Here Is My Right," as one of unity and inclusion: "Working together teaches tolerance for other voices and their acceptance. Curiosity about other cultures helps to raze the walls inside one's head. The HipHop Culture forms the ***background*** for a protest movement, originating from the street. . . . From a violence-free and anti-racist union comes a youth culture scene, concrete works of which are achieved in a realistic periphery—and, that periphery can only be realistic if one carries out his/her daily life there. Think globally—Act Locally."

During the festival, various rap and graffiti workshops took place, attracting renowned local street artists and numerous rappers and writers-to-be. Also present were prominent graffiti writers from other metropolitan centers, such as T-Kid from New York and Jay-One from Paris. They were in Berlin as invited artists and conversed with their Berliner hip-hop brothers on aesthetics of graffiti and on the ethics of hip-hop. In a panel discussion on style with T-Kid and Jay-One, the hip-hop community was heralded as a multicultural community, extending beyond borders and delivering the message of peace and brotherhood. In their interpretation of the "imagined community" (Anderson 1991) of hip-hop, graffiti on the walls were messages addressed to society at large, as well as being artistic expressions of individual writers. In this world, they claimed, styles were many and varied, and so were the cultures and writers.

The hip-hop boys and girls of Berlin were indeed innovators of style and meaning. They aspired to be street writers, preached rhyming against violence, and espoused tolerance and multiculturalism. Their enactments of hip-hop were connected to the global hip-hop scene in tangible ways, from clothes they were wearing (baseball caps, baggy jeans) and the vocabulary they employed (respect, street, brotherhood). But their enactments of hip-hop community were not simply imitations of the ghetto bravado of other places. They also saw themselves as part of the Europe that is emerging. In Europe, as one writer argued, a new style, a "European style," certainly different from the American style, was developing—a style dominated by "smoother lines and figurative drawings" as opposed to the "geometric designs" of earlier times and other places. More importantly, the imaginative tag of their posse, "To Stay Here is My Right," was an intervention in civic spaces. As it was proclaimed in the posters of Street '94, this right was not bounded by timeless identities, as in Germanness or Turkishness; it was an undifferentiated ***right to stay in Berlin,*** whoever they were and from wherever they were, thus disrupting the normative, nationally marked conventions of the debate about migrant youths.

This universalistic claim, and the utopian—one could also say fictional—world of hip-hop, was self-evident in the Kreuzberg of 1994 and owed its possibility to the organizing discourses of pluralism and tolerance in which Berlin's metropolitan self-presentation was entrenched. While the festival of Street '94 was celebrating cosmopolitan pluralism, Berlin's Foreigners' Office had been promoting the diversity of a city in the making. In 1992, for instance, a poster published by the office covered the billboards of the city and pronounced "Wir sind Berlin" (We Are Berlin). The Berlin in the poster comprised the portrait photos of persons of diverse professions, ages, colors, and genders, without any identifiable reference to the nationalities or ethnicities of the persons. Later in the 1990s, the Foreigners' Office launched another poster campaign called "Was ist deutsch?" (What Is German?), listing numerous stereotypical images of Germanness ("Bureaucracy? Garden dwarfs? Sauerkraut? Uniforms? Humor? Beer?").

Posed against the larger question "What Is German," these images were intended to problematize Germanness for the sake of plurality. In addition to "We Are Berlin" and "What Is German," the Foreigners' Office was (and still is) involved in other campaigns, some successful, some utterly misguided (posters of love between a migrant black girl and a neo-Nazi youth, for instance), all in the service of achieving a cosmopolitan Berlin, tolerant and diverse. Street '94 and the To Stay Here is My Right posse clearly belonged to this new imaginary of Berlin.[9]

■ From the Kreuzberg onto European Stages

Following the success of Street '94, the youths of NaunynRitze staged another megaproject in 1995: Multinationales Anti-rassistiches Performance Project (MAPP). Like Street '94, MAPP was designed to actuate the productive potential of the youths. With this project, however, NaunynRitze was shifting its focus from the ghetto scenes of Kreuzberg to European stages. The project was part of the European Council Initiative Campaign on Combating Racism, Xenophobia, Anti-Semitism and Intolerance and was designed to bring together European youths in multicultural and antiracist performances. The funds for the project came primarily from the European Union's Youth for Europe program, with supplementary funds from the local governments of the cities involved. It was a joint production with four other youth organizations from Sheffield, Rotterdam, and Luxembourg—the Sheffield Youth Center, Centrum de Heuvel, and Maison des Jeunes et de la Culture and Maison des Jeunes, respectively.

The Youth for Europe Program and its sister program, European Voluntary Service (EVS), commenced in the mid-1990s under auspices of the European Commission's Directorate General for Education and Culture, which is the body responsible for designing and carrying out policy in the fields of education, training, and youth. Youth for Europe was launched in 1995 as a five-year project with an allocated budget of 126 million ecus (European Currency Unit) ($113.4 million), whereas EVS was carried out as a pilot project between 1996 and 1997. Both Youth for Europe and EVS targeted youths between the ages of fifteen and twenty-five, with a specific mission, promoting cultural exchange between the youths from fifteen member states of the EU and countries belonging to the European Economic Area. In 1995, more than 6,000 youths from eligible countries in Europe participated in one thousand projects sponsored by the Youth for Europe program. In 1996, for instance, about 65,000 youths took part in four thousand projects under the Youth for Europe program. During its pilot run, the EVS program mobilized about 2,500 youths to carry out voluntary local work in various member countries. Following the success of the pilot, the EVS program became fully operational in 1998. Currently, both programs are merged with the EU's newly endowed Youth Program, which continues to sponsor projects and youth exchanges. Whereas in the beginning their area of action was limited to the EU member countries, later the programs were expanded to cover pre-accession countries and non-EU "third countries," particularly ex-Soviet republics and countries in Africa, the Mediterranean, and Latin America.[10]

As stated in its official declaration, the primary objective of the Youth for Europe program is to direct EU resources to providing information to and building networks among the youth. Through this program, the EU aims to "[address] young people outside formal education and training structures" and to "[widen] the possibilities offered to young people to take [*sic*] contact with Europe and to participate in its construction as active and responsible citizens." To achieve this goal, the EU sponsors cultural projects, which aim to further the goal of Europe "as an integral part of the historical, political, cultural and social environment" of young people, "[promote] an awareness of the dangers relating to exclusion, including racism and xenophobia," and "[encourage] young people to find out about, become aware of, and recognize the intrinsic value of cultural diversity." Many youth centers engage in projects that subscribe to these goals and receive funds to carry them out. MAPP was one such project among many, receiving 142,500 ecus ($128,250) from the program.

In the summer of 1995, the show *Against Racism—Culture on Tour (Gegen Rassismus—Auf Kultur Tour),* traveled to all four participant cities, with a cast of sixty youths and their social workers. According to the social workers involved in the project, the tour was the high point of collaborative work, which had started in 1993. After five days of joint rehearsals in Luxembourg, the MAPP company staged their show ten times in front of multicultural audiences—estimated to be a total of five thousand youths in all four cities. The tour was named differently in each country, keeping a thematic commonality and attention to hip. In England it was called *Challenging Racism through Performance and Participation;* in the Netherlands, *Jongeren in Aktie Tegen Racisme door Middel van Theater, Muziek en Dans;* and in Luxembourg, *"Cool-Tour": Géint Rassismus.*

The show was a multigenre performance, a mix of dance, theater, graffiti, acrobatics, and music. Its themes, antiracism, diversity, and tolerance, were set as societal problems as much as problems facing the youths at large. The brief leaflet distributed by the MAPP company before the show in Berlin addressed the audiences in a voice of social responsibility and explained the purpose of the project. In the company's own words, "In a collaborative environment, they [were] experiencing themselves through a learning process, involving intellectual, social, and emotional exchange. Through their show, [they were] seeking to achieve communication in order to prepare themselves to revolt against hate directed at foreigners, to be ready to have respect for learning from others, and to learn that only with cooperation of all the youths, peace, solidarity, and togetherness can be achieved." As envisioned by its performers and producers, MAPP was an investigation of "intolerance and boundary-making in abstract, concrete, and symbolic ways," and its story was "based on a classic story of two lovers, whose love becomes impossible because of their different origins." This story of unrequited love between two youths from different ethnic backgrounds, billed as a popular remaking of Romeo and Juliet in ethnic visage, won NaunynRitze a Mete Ekşi Prize, given in memory of a Turkish youth killed in a street fight. When presenting the award, Barbara John, Berlin's commissioner for foreigners' affairs, commended the youths for their antiviolence message and their dedication to the cause of living together: "With this Prize, Mete will be remembered as a young man who lived in Berlin and embraced the idea of peaceful living. You, the winners of this year's Prize, now share the responsibility of furthering Mete's ideas."[11]

At one level, with its easily recognizable, formulaic story line of youth love and rebellion, MAPP was yet another youthful production, variants of which were (and have been) abundantly performed in Berlin's public spaces year after year. These performances are fixtures of Berlin's informal stages (schools, youth centers) for youth art and activism. They follow and exhaust local fashions and global trends. Rappers compete with rock 'n' rollers to capture the seal of cool; techno-kids steal the show to become the next big thing. At another level, however, MAPP's relevance exceeds the modest and episodic place it occupies in the ever changing world of youth art. With MAPP, the migrant youths of NaunynRitze were (and have been) connecting Kreuzberg to Europe in (in)tangible ways. MAPP linked Kreuzberg to Sheffield, Rotterdam, and Luxembourg on a map of Europe and Europeanness by way of elaborate bureaucratic exercises and perpetual invocations of European principles of antiracism and diversity. Writing grant applications, competing for European funds, receiving awards for intercultural understanding, traveling and performing as members of multicultural troupes, for instance, were acts and enactments, aggregates of which, albeit implicitly, facilitated the spatialization of Kreuzberg as European space or the space of Europeanness.

Migrant Youth, Berlin, and Europe

Speaking of Berlin, the reunified cosmopolis at the beginning of a new millennium, we picture cityscapes and urban masquerades and converse about projecting futures and forging transnational connections. In and around Potsdamer Platz, where the Wall once divided the two Berlins, we see the largest construction site in the world and envision a Hauptstadt, Kulturstadt, or Weltstadt in the making. In our narratives, a captivating Berlin welcomes its residents and visitors to a place of diversity and cosmopolitanism for the new Europe—even though our captivation with the place is afflicted by disconcerting episodes of its past and present.

In this Berlin of projected diversity and cosmopolitanism, when it comes to migrant youth, we still prefer to discover timeless traditions and unbridgeable cultures. We carry the burden of seeing *natural* and *exotic* formations in the making. In our scholarly and everyday conversations and policy debates, youth appear in tales of the second generation and integration—disoriented on street corners and disconnected from the larger society, its institutions, resources, and discourses, standing at an incommensurable distance from the modernity and present tense of the new Europe.

When we see young Turkish rappers and graffiti writers in baggy clothes or Turkish girls donning the latest fashions or wearing headscarves, we conveniently forget the social settings within which they enact identities and assert claims. We disregard the place of contemporary discourses of tolerance, equality, and diversity in the language and content of their claims. They are, however, active participants in the societal projects of their environs and times and cultivate youthful imaginations. They are inhabitants of Berlin, and the cultural projects they are engaged in contribute to the constitution and imaginary of the city as a diverse, cosmopolitan metropolis in the new Europe.

The episodic displays of youth culture and politics that I sampled in this chapter are not singularly representative of the orientations and political aspirations of Turkish youths in Berlin. I contend, however, that through these episodic productions, conversations, and dreams, migrant youths speak of their conditions, expectations, and resolutions in the public spaces of Berlin, and they speak to the world at large, articulating utopias against, and because of, the uncertain eventualities encompassing their lives. While the vibrant performance of hip-hop on the stages of Street '94 was a means to converse about universal brotherhood and unconditional right to stay, the rather didactic proclamations of MAPP on racism and diversity were calls for solidarity to fight social problems evident in Europe. Whereas the youths of Street '94 were sending utopian messages from Kreuzberg to the world at large, the performers of MAPP occupied European stages and added their voice to Europeanness. Through these enactments of youthful resistance, Turkish migrant youths of Berlin perform cultural diversity for the metropolis and participate in the project of Open City Berlin, Europe, that which is unfolding but not foretold.

NOTES

1. The excerpts in *Die Zeit* are from Schmidt 2000. My citations, here and in the next few paragraphs, are from *Die Zeit*'s Web edition published at www3.zeit.de/2000/41/Politik.

2. The approach advanced here draws from Castells 1998; Sassen 1996, 1998; Soysal Nuhoğlu 1994, 1997, 2002; Varenne 1993. The scholarly work on Europe, however, is broader than the works that inform my analysis. For consequential examples from the field of sociology and political science, see Delanty 1995; Garcia 1993; Kohli 2000; Smith 1992, 1993; Therborn 2001; Weiler 1999a. For studies of Europe from anthropological perspectives, see Borneman and Fowler 1997; Goddard, Llobera, and Shore 1994; Herzfeld 1987; Pina-Cabral and Campbell 1992; Shore 2000; Modood and Werbner 1997; Wilson and Donnan 1998; Wilson and Smith 1993. For an extensive review of the formal relations of Turkey with Europe and the European Union, see Balkır and Williams 1993; Müftüler-Bac 1997.

3. In the cultural studies canon, *liminality* is commonly read and valorized as a positive space—a space of becoming and resistance (see, e.g., Bhabha 1994a, 1994b). Nonetheless, liminality—thus the migrants—occupies a space outside home and host countries, reifying the territoriality of migrations and identities.

4. The project of Open City Berlin is pursued by a consortium of governmental and private agencies in association with the Berlin Senate, among them Partner für

Berlin (Partnership for Berlin), Architektenkammer Berlin (Berlin Chamber of Architects), the Land Office for the Preservation of Historical Monuments and Buildings, the Senate Department of Urban Development, Environmental Protection, and Technology, the Senate Department of Construction, Housing, and Transport, and the Federal Ministry of Transport, Construction, and Housing.

5. In Europe, Germany has the largest migrant population from Turkey, currently around 2 million; and in Germany, among metropolitan centers, Berlin has the largest number of migrant residents.

6. Like Kreuzberg, Wedding, Schöneberg, Tiergarten, and Neukölln have high percentages of Turks and other migrants living within their boundaries. In terms of sheer population statistics, these districts can also be identified as centers of Turkish presence, but they are considered to lack the touch of "diversity and hip." For more detailed population statistics for Berlin, see the city's Web site at www.berlin.de.

7. The argumentation presented here takes its cues from, and expands on, the body of critical commentaries and interventions advanced in the following two sets of studies on youth and culture: in the area of (sub)cultural studies and youth, Amit-Talai and Wulff 1995; Baumann 1996; Cohen 1988; Gilroy 1993; Griffin 1993; Hall and Jefferson 1983; Hebdige 1979; Pilkington 1994; Willis 1977, 1990; in the domain of critical examination of cultural approaches in social sciences, Abu-Lughod 1991; Fabian 1983; Gupta and Ferguson 1992; Herzfeld 1987; Malkki 1992; Moore 1989; Soysal Nuhoğlu 1994; Stolcke 1995; Wallman 1978. For a sample of studies on Turks in Germany, with varying degrees of emphasis on—and critical stance against—culture, see Çağlar 1997; Horrocks and Kolinsky 1996; Mandel 1996; Schiffauer 1991; White 1997; Wolbert 1995. For literature specifically on Turkish youth, see Bröskamp 1994; Çağlar 1998; Heitmeyer, Müller, and Schröder 1997; Kaya 2001; Nohl 1996; Soysal 1999, 2001; Tertilt 1996.

8. The data presented here draw upon my research in Berlin on youth culture and formations of identity among migrant youth groups (between 1990 and 1996, I spent three years in Berlin for my fieldwork). My research primarily focuses on Turkish youth organized in various cultural and political associations and attending cultural/recreational centers, and it explicates and reflects upon the experiences of migrant youth in such organizational settings. The outcome of this research, an ethnographic analysis of migrant youth culture in Berlin, is compiled in Soysal 1999.

9. On the pluralist policies and efforts of Berlin's Foreigners' Office, see also Çağlar 1998; Soysal Nuhoğlu 1994; Vertovec 1996.

10. For current information on the EU's youth programs, see the Web site "Education, Training, and Youth" at europa.eu.int/pol/educ/index_en.htm. An evaluation of the Youth for Europe and EVS programs are available at the site. Note that the Web site and its contents change as the EU policies and policy instruments are modified or altered. The data presented here include citations and statistics posted on the Web site in 1998 and 1999, hard copies of which are in author's personal archives. Although the EU has reorganized its youth programs, there have not been major shifts in policy and policy instruments since 1995.

11. *BLZ*, January 1996, no. 1, pp. 20–22.

PART IV

Democracy and Identity

NINE

Territoriality and Political Identity in Europe

JOHN AGNEW

National identities are often seen as both prior to and the polar opposites of political identities associated with other possible territorial scales, such as the local or the regional, on which political identities, or politicized social identifications, are formed. Under contemporary circumstances, however, including the globalization of economies, the questioning of the efficacy of existing states, and the increased integration of many European states, national identities may no longer be best thought of as either superordinate or opposed to political identities on other territorial scales. Using the case of the Northern League in Italy, this chapter identifies *three new "rules" of political-identity formation:* the self-conscious invention of political units, the malleability of identities, and the multiplicity of identities; these rules suggest the mutual contingency of different territorial scales in the crafting of political, including national, identities.

First, however, I want to dispute two critical assumptions of conventional views of the relationship between territory and identity. These assumptions are common to international law and institutions as well as to many strands of contemporary social science and must be dealt with before an alternative to them can be proposed. The first assumption is the homology that is drawn between individual persons and territorial states such that *states are treated as if they were the moral equivalents to individual persons.* This both familiarizes the state and gives it a moral-political status equivalent to that of a person. This assumption privileges the scale of the territorial state, and hence its primordial role in political-identity formation, by associating it with the character and moral agency of the individual person, an intellectually powerful feature of Western political theory. In medieval European political thought, the state did not have such an exalted status, a fact that points to the historicity of the relationship between a singular territorial scale—that of the state—and political identity. The second critical assumption is the dyadic (person-person, person-state, state-state) definition of ***the nature of social relationships upon which the territory–political identity relationship rests.*** This reduction abstracts political identity from the multiscalar sociological contexts in which it originates and operates into a set of isolated individual relationships (particularly that between state and individual) on singular territorial scales. In this way political identities are taken as given categories of belonging associated with mutually exclusive group memberships bounded by specific territories without examining the processes that actually produce the categorizations upon which the identities and their territories are based.

■ Orthodox Assumptions about Territoriality and Political Identity

The Moral Geography of the State and Political Identity

Exalting the state as the singular font of political identity has involved equating the state with the apparent autonomous identity of the individual person. This is not to say that there is no analogy between the *social construction* of personhood and that of statehood. Rather, it is to deny the atomized understanding of both states and persons that conventional approaches to personal and state agency entail. Not only is this a desocialized view of the person or the state, implying an essentially transcendental persona making itself; it also turns sovereign states into naturalized abstract individuals that can then be inscribed with the moral authority of their own personhood. Modern statehood is thus underwritten by modern personal individualism. A moral argument about the equating of the autonomy of the individual person with that of the state is masked by the ***natural*** claim that is made on behalf of the state as an ***equivalent*** individual.

The appeal of this strategy has been twofold and reflects a powerful aspect of both the modern social construction of statehood and the privileging of national identity in both political practice and analysis. First, it allows, as Thomas Hobbes was perhaps the first to note, identification of a historical "state of nature" in which a set of primitive individuals, liberated from social conditions, can compare their natural condition with that offered by a specific set of social-political conditions associated with statehood (Jacobson 1998; Skinner 1999). Actual power relationships can thus be ascribed to the need for security or wealth in a world that is not the independent creation of any single individual (Held 1995, 38). The separation and isolation of individuals (persons, states) thus produces a logical case for a pooling of power in the hands of a single sovereign. It is now what can be called the Standard American Social Science Model, in which a ruleless state of nature is assumed from which socially disciplined individuals emerge by means of state conditioning (see, e.g. [across the political spectrum], Buchanan and Tullock 1962; Nozick 1974; Rawls 1971; and Sunstein 1993).

Second, the unrelenting suspicion and hostility with which human individuals regard one another in Hobbes's state of nature and, paradoxically, the humanistic tendency to raise the self-aggrandizing individual onto an intellectual and moral pedestal in much Western thought (drawing from Adam Smith) underpin the projection of the qualities of personhood onto statehood. At one and the same time, the state embodies the two sides of the modern moral coin. The state represents (pace Hobbes) the territorial solution to human aggression and fear of untimely death among a discrete group of persons, displacing aggression into the realm of interstate relations. The state is also constructed as a primitive individual such as a person with unique abilities, particularly its ability to specialize within a division of labor (pace Smith), which potentially offers a way of maximizing output and thereby increasing the wealth and satisfaction of all. These two moves, the one political and addressing human aggression and the other economic and addressing human acquisitiveness, boost the state into a position of historical preeminence in relation to the definition of political identity irrespective of their empirical veracity (Inayatullah and Rupert 1994).

Yet, even though I acknowledge the powerful influence of this equating of the state and the individual person on actual practices of statehood down through the years, there are ways in which its contemporary adequacy can be questioned on empirical as much as theoretical grounds. One is to point to the lack of unity in the making of foreign policy by states. This involves pointing out the distinctive positions adopted by different sectoral and geographical interests within states.[1] Different social and economic groups, lower-tier governments, and sectional lobbies all bring different identities and interests to bear that under certain circumstances can burst into civil wars or immobilize state institutions. In addition, statehood is the result of mutual recognition among states and not the result of isolated states achieving statehood separately and then engaging with one another as abstract individuals, in the sense of thoroughly autonomous persons (on the first point, see Biersteker and Weber 1996; on the second, see Lukes 1973, 46–47, 76–77, 99). The importance of the Treaty of Westphalia, for example, lay in its legitimation of the emerging territorial states as neutral centers of public power imposing order on warring religious factions (Hirst 1997, 216–35).

Another less frequently chosen route to pointing out the empirical contingency of statehood as thought of as morally equivalent to personhood is to argue historically by referring to the social-moral position of statehood in medieval political thought, thereby demonstrating through its difference from modern conceptions the historical contingency of the mode of thought we now tend to project backward and forward in history. The state as a unitary actor akin to a modern person is the product of the determinative social conditions of European modernity. This point is worthy of further comment. As Monahan (1994, 8) has noted, medieval writings offer a good opportunity to "address our own questions to the past so that it can teach us."

Long before Hobbes, medieval European political thought had arrived at a quite different interpretation of statehood from the one that the later masters such as Hobbes were to offer. Though obviously more complex than I give credit for here, medieval thought tended to work down from an ideal of human unity within a God-ruled universe to a "horizontal" world in which an "antique-modern" conception of the state as a centralizing and territorializing power (drawing from the ancients and some contemporary trends toward monarchical absolutism) vied with multiple loyalties to spiritual and temporal authorities and overlapping jurisdictions as the dominant normative standard for judging political rule (see Nederman and Forhan 1993). What is clear, however, at least from the classic account of Otto von Gierke (1922, 70–72), is that initially the state had no separate personality. Only the "visible wielders" of power or a "Ruler-personality" was recognized as having authority. People were merely the collective sum of all persons, not a collective entity in its own right (that this may be something of an exaggeration, however, is the conclusion of Reynolds 1984). As a result, "the path to the idea of 'State-sovereignty' was barred for medieval theory" (Gierke 1922, 73). Indeed, for many years there was widespread acceptance of the idea of a hierarchy of communities with specified purposes and overlapping spatial jurisdictions, which was only slowly replaced by that of "a theoretical concentration of right and power in the highest and widest group on the one hand and the individual man, on the other, at the cost of all intermediate groups" (87). In such a world, no single territorial scale monopolized the production of political identity.

Though the rise of the modern territorial state obviously did lead to its increasing monopolization of the processes of political-identity formation associated with the state's increased centrality to people's everyday lives in providing infrastructure, regulating social and economic affairs, defining citizenship, and categorizing populations, traditions and memories of common history had to be constantly invented and reaffirmed in order for national identities to form and persist across generations, as we are reminded by Hobsbawm and Ranger (1983). National identities never just happen. Their intellectual and popular strength has been particularly reliant on the privileged status accorded to the modern territorial state as a moral equivalent of the individual person with its own identity and interests. It is only as statehood has been brought into question by the failures of states to live up to what they promise their populations that the moral geography of statehood has also come into question.

The Geosociology of Political Identities

If the moral geography of statehood subordinates the individual person to the "personhood" of the state, then the individuation of persons makes them available for subordination by the state. A geosociology of political-identity construction, however, suggests otherwise.

In the first place, personal political identities and their associated territorial scales are not pregiven. We should start with the flow of events and actions in social and political life out of which contingent political identities are formed rather than presupposing fixed and stable identities, however commonsensical and convenient such designations might seem. Consider, for example, the history of the British political identity, emerging in the eighteenth century, strengthening during the nineteenth and early twentieth centuries, but now enduring a marked regression in the face of assertive Scottish, Irish, Welsh, and English political identities.[2] Contingent identities emerge to gain a footing or control in an uncertain world, where goals or intentions are defined by presuming the intentions of others, in which errors in interpretation and judgment feed back into identity, and biographies or histories are told and written to clean up the identity and make it self-evident. These are the steps to "personhood" identified by Harrison White (1992) in his social theory of identity. In this construction, personhood develops out of identity struggles for control in a mix of social networks in which putative persons are enveloped from childhood. White's main point is that "stable identities are difficult to build; they are achieved only in some social contexts, ***they are not pre-given analytic foci***" (201, my emphasis). A large, geographically encompassing social context is required for political identities to develop. In other words, an identity is not ontologically prior to a set of intersubjective relations. An identity is defined and recognized as such only within a set of social relationships that establish rules for what is and what is not acceptable as one: the process of identification, if you will.[3] An identity, therefore, is never simply the outcome of action and imposition on a single geographical scale, such as that of individual territorial states, but also of necessary social rules working on more local and also on broader territorial scales. Just as identity is not pregiven, neither is a territorial scale on which an identity is defined and with which it is associated.

The labels used to identify territorial scales—or territorial units of different sizes or geographical scope—are, of course, intellectual (and social) constructs rather than simply existing "out there," so to speak. Though they suggest a mutual independence, almost discrete slices of reality, in practice each scale—local, regional, national, international, global—implies the existence and impact of the others. An order of listing also indicates the implicit relative valuation of each in any particular account that makes use of that listing. Moreover, it is not the separate scales so much as the putative relations between them that help define the social context for the formation of political identities. The balance between social processes (of power, status, and discrimination) operating from sites with differential geographical scope determines the ways in which identities come about and are deployed (Howitt 1993, 1998).

Second, political identities emerge within networks of subjectivity as a result of both quantitative capacity or conventional power over ***and*** power in the sense of the ability to bind others into networks of assent. In this understanding it is not the ability of a single actor (such as a state) to create assent that is privileged but rather ***the strength of association between actors based on shared norms and values.***

This would include, but not be restricted to, the writing of agendas, the silencing of certain options, and other modes of "mobilization of bias." It necessarily covers, therefore, the "rules of the game" among actors. Historically, these rules have emphasized "hard" or coercive power, particularly that exercised by states. Today, some commentators argue, they involve the much more pervasive use of "soft" or co-optational power, in which assent has become more significant than coercion (Allen 1998). This reflects the emergence of a world in which diffuse economic transactions and *chosen* interpersonal relations are more vital to the constitution of political identities than is coercion. But, by definition, even coercion requires the application of commonly accepted rules of conduct concerning its purposes and limits that all involved must at least tacitly or subconsciously acknowledge.

Third, and finally, the social networks in which political identities are *embedded* have historically defined geographical settings. In nineteenth- and early twentieth-century Western industrial society, for example, networks were confined within rigidly bounded state-territorial, imperial, and world-regional settings. Jeremy Bentham's understanding of the power of the regulatory gaze exerted through state bureaucracies exemplifies the epoch, as Michel Foucault has famously claimed. A rigid territorial core-periphery model ruled identity construction. In Timothy Luke's terms: "relatively clear relational-spatial distinctions of social status, cultural preeminence, and political authority develop[ed] in line with a print-bound, panoptical space in traditional industrial societies or industrializing agricultural societies" (1989, 47). In other words, raw material–manufacturing linkages within territorial empires, print-based media of communication, and centralized state apparatuses produced a territorialized set of relationships between local nodes of networks with denser connectivity within state boundaries and more attenuated links across the globe.

In contemporary "informational society," however, the spatial-temporal character of political-identity formation is being transformed. Luke describes this trend:

> With the growing hegemony of transnational corporate capital, the means of information become the critical force in modern modes of production. A new politics of image, in which the authoritative allocation of values and sanctions turns on the coding and decoding of widely circulating images by politicized "issue groups," arises

> alongside and above the interest-group politics of industrial society. Contesting these mythologies can expose some of the contradictions and hidden dimensions of image-driven power. But, on the whole, the endless streams of mythological images in turn bring together the flow of elite control, mass acceptance, and individual consent in a new informational social formation—the "society of the spectacle." (1989, 51)

This is a relatively deterritorialized network system in which nodes are widely scattered around the world; the nodes are more densely connected within and between Europe, North America, and East Asia, though still constrained by territorial structures inherited from the earlier epoch (Harvey 1989a; Scott 1998). The geographical embeddedness of political identities today, therefore, signifies something different from what was the case in the past in terms of the relative balance between territorial and geographical-network elements and the geographical scope conditions (broadly, from national-international to local-global) under which political identities are produced.

■ Myths of Territory and Identity in Contemporary Europe

A critical social geography of territory and political identity such as that laid out in the previous section has rarely prevailed in modern social science. I would identify several dimensions of how considerations of political identity have been related explicitly and implicitly to understandings of territory. These have taken on the status of myths guiding the practice of thinking about territory and political identity. One of the commonplaces of mid–twentieth century social science has been that territories or places (the spaces of various sizes occupied and lived in by people) other than that of the container-state have no *independent* role in political identity in the modern world. Class, religion, ideology, and gender are variously seen as primary. Even nationalism is deterritorialized as an ideological projection of the imagination of groups that make no essential claims to specific bits of territory, only to "imagined communities."[4] Assuming that the existing boundaries between states define something akin to a territorial state of nature (as is common in political theory) allows discussion of political identity to ignore the geographical parameters of who is in and who is outside of a particular political project (on this in relation to Italy, see, e.g., Allen and Russo 1997). Social scientists simply presume that debate about and actual contestation of political identities takes place within the territorial parameters of existing state-based territorial structures.

This is not to say that this presumption is invariably false. Indeed, a case can be made that in Europe from the nineteenth century until the 1960s, the primary context of political-identity formation was largely that of the established nation-states. In recent years, however, this has changed, mainly as a result of increased international communication, the development of the idea of an integrated Europe, the strengthening of regional economic disparities, and the flowering of regionalist movements. Today, a hierarchy of potential territorial affiliations presents itself to the average citizen from Europe (or the world as a whole) through existing nation-states to regional nations (such as Catalonia, Scotland, and Corsica) and localisms. Of course, identities can form only if there are stories or narratives about them available for popular use (Somers 1994). Social cleavages and economic dependences cut across territorial differences in complex ways to create multiple identities that have a range of intrinsic geographical definitions. For example, with respect to language, a whole new cultural configuration is under construction in Europe which is more like that of India since 1947 than France after 1870. The emergence of regional languages and languages of wider currency, particularly English, portends a normalization of multicultural identities (perhaps even a Europe-wide one) rather than the establishment of mutually exclusive linguistic realms in which a single linguistic and associated political identity will prevail (Laitin 1997).

A second dimension of conventional wisdom about territory and identity is that to the extent that there are acknowledgeable local and other non-national-scale cultural and political differences, they are seen as products of the long-distant past rather than as constituted over time and potentially invented in the present. The dominant stories told about Italy, for example, by scholars as well as politicians, are about how the north and south are politically different today because of either (1) the forms of government they had in the Middle Ages or (2) the historical ***amoral familism*** (or ethics that extended only to family members) in the south as opposed to the north; the stories assert that there is continuing powerful localism because Italy unified late (in the 1860s) or has experienced a peculiar kind of industrialization dependent upon a familial capitalism of a few large firms and millions of very small ones established in the distant past (Agnew 2002, chap. 4). There is little or no place for considering the role of the contemporary institutional framework, the national economic structure as constituted since the 1960s, and the operations of political parties since the Second World War in not so much reproducing as *creating* the geographical differences that are so much a feature of contemporary Italy. To the extent that generalizing about Italy remains problematic, therefore, commentators have a stock of stories available in which present-day geographical differences are put down to past social and political ones (e.g. Putnam 1993).

Finally, to the extent that one geographical scale is seen as emergent and threatening to the hegemony of the national, it is the local one. This is very much the sense that one derives from the burgeoning literature on globalization, particularly in the fields of cultural studies and political sociology. Though rarely defined in other than the vaguest terms, the local is often sited at one end of a global-local dialectic to define a world in which the national scale, on the one hand, and geographical scales other than the local and global, on the other hand, are in eclipse. Under conditions of globalization, as localities are inserted directly into the networks of global capitalism with diminished national state regulation, the local is seen as replacing or substituting for the national in the reproduction of uneven development. A spatial dialectic such as this only has room for two geographical scales. To the extent, therefore, that a political identity is in crisis, it is seen as the national; and to the extent that a new scale is in the ascendancy, it is the local (which usually seems to subsume the regional). In this view, political identities are invariably substitute rather than complementary goods, and the local/regional is seen as increasingly substituting for the national. Yet, as Alon Confino (1997, 23) shows persuasively for Württemberg, Germany, in the period 1870–1918, national identity is not necessarily exclusive of other identities; rather, it is constructed between the "intimate, immediate, and real local place and the distant, abstract, and not-less-real national world." However, in the contemporary context of globalization—the unmediated insertion of localities and regions into world markets and the emergence of higher-tier governments such as the European Union—it is not simply the local and the existing national that are involved. Instead, the current trend is for a renegotiation of the relationships between various scales of political identity, including the national. As a result, a multiplicity of crosscutting territorial identities—local, regional, national, and so on—are available for stimulus and mobilization by political movements and populist politicians.

The Northern League and Political Identity in Contemporary Italy

One of the most seemingly peculiar yet intriguing political phenomena in contemporary Europe is the Northern League in Italy. Originating in a number of localistic movements in the early 1980s, the League took on its present organizational form in 1992. Its leader, Umberto Bossi, has shifted back and forth between demanding a new federal Italy and proclaiming outright secession (Agnew 1995). Now the major or second party in both national and local elections in northern Lombardy and the Veneto region of the northeast, the League currently stands for an independent Padania, a new state, variously defined but usually claiming Italy from its northern alpine boundary as far south as a line running along the southern borders of the administrative regions of Tuscany, Umbria, and the Marches. For those of us bewitched by the twists and turns of Italian politics, the League, as an explicitly northern-regional party, represents a novel departure from the nationally oriented if rarely nationally successful character of the two major postwar Italian political movements. The ideologically universalistic Communists dominated central Italy, and the equally (if distinctively) universalistic Catholics (in the Christian Democratic party) dominated the northeast and parts of the south. Variously explained as a protest movement or an antisystem party whose future survival has been seen as unlikely by most orthodox political scientists, the League has had a remarkable staying power over the past fifteen years (Cartocci 1994). Part of this relates to the fertile period in which it was planted and grew, as the collapse of the old system of parties in 1992 — in reaction to the great corruption scandal of *tangentopoli* — provided an opening to new parties running against the old system, its politicians, and their model of national integration.

But it is my contention that the success of the League also indicates in an extreme and focused form three lessons that can be drawn from its experience that may well say something more general about a developing aspect of the politics of identity in Europe and North America: first, the explicit and self-conscious *design* of new territorial entities — in this case, known as Padania — where none had much more than the slightest existence before; second, the *malleability* of political identities after a period in which these had seemed to take on a permanent cast, particularly in terms of the priority of established national identities over those associated with other geographical scales and in terms of the emergence of geographically differentiated class, gender, and religious identities that were supposedly invariant within existing national territories; and third, the *coexistence* of multiple political identities, no one of them necessarily replacing all others. These three lessons about territory and political identity and their manifestation in the Northern League are what I want to address in this chapter.

The Northern League and the Contemporary Politics of Political Identity

It is in this theoretical context that the case of the Italian Northern League is interesting and important. From the League's recent experience, as interpreted by means of my fieldwork in Varese and elsewhere in northern Italy and the results of recent Italy-wide and northern opinion polls, I want to emphasize three new rules relating to political-identity formation that have consequences for theoretical debates about place and political identity. I chose Varese as the setting for interviewing League politicians because most of the important leaders of the League, including Bossi, come from the province, because the League has become the main electoral force in the province since 1992, and because the League has been strongly established in local government in Varese since the early 1990s.

Designing Padania

The first of the three rules involves the self-conscious invention of a new political entity, the Italian north or Padania, complete with its own myths, symbols, and rendering of history. Even though the League's leaders largely accept the view of pollsters that it is local, not regional, identities that have lain behind much of the support for the movement, they have come to believe that to achieve success in decentralizing power from Rome, they must create a new territorial entity upon which they can focus their aspirations. Having failed in a previous strategy of wresting power from Rome by participating in Roman (national) government (after the 1994 election) to promote some type of federalism, by 1995 they were investing their hopes in welding together an "imagined community" for northern Italy as an alternative to Italy as a whole. The deliberate nature of this process shows how much has changed from the last great round of creating national units in the late nineteenth century, when nations were not so much designed or customized as they were off-the-shelf products of at least minimal ethnic distinctions. All nations may have founding moments, but these are usually the outcome of long periods of prior excitement, enthusiasm, and mobilization in which the symbols of nationhood, not least the claimed national boundaries, arise from widespread acceptance of their naturalness and rootedness in a common past, however mythic that past might be. If, as Anthony Smith (1991, 100) contends, the element of "design" was never "wholly absent even in the English, and later British, case," the nation in this and other Western cases, if not in the world liberated from European colonialism, was never invented from scratch. Preexisting "ethnic configurations" were vital in the formation of nations. The Northern League, therefore, may be the first authentic postmodernist territorial political movement in its *self-conscious* manipulation of territorial imagery to create a sense of cultural-economic difference from an existing Western state of which it is part. Italy, of course, as League activists never tire of telling interviewers, was always the least natural of cultural units for European nationhood. Ironically, the nineteenth- and early-twentieth-century attempts at unifying Italy under Liberal and Fascist regimes faced a continuing dilemma of creating Italians out of a disparate populace—a task not dissimilar to the one facing the League with Padania. The difference, however, is that there was a geographic entity called Italy that did have some prior history of political and economic unity, and there was no sudden, deliberate concoction of a territorial entity upon which political hopes could be projected, such as has occurred with Bossi's invention of Padania. Italy may have been only a "geographical expression" in 1860, but it was at least that.

The League has had few if any primordial ethnic elements to play with, except for a set of dialects, many of which are mutually unintelligible. The region and its history have had to be made from scratch, largely in terms of giving cultural dressing to resentments about the mismatch between the contemporary institutional structure of Italy and the economic trajectory of its northern part. A geographical entity thus provides the basis for making a set of historical claims rather than a set of historical claims providing the basis for making a geographical claim. This is a powerful example of how "lived space" is perhaps becoming the "everything" that Edward Soja (1989) has claimed for it. First of all, Padania has had to be given a territorial definition. The League has felt the need to reify its cultural claims about northern cultural distinctiveness by giving the region definite borders, even when these seem to go well beyond the geographical limits of mass support or the understanding of militants, who have tended to have a more geographically constrained definition of northern Italy than does Bossi. From the League's perspective, however, these borders do not involve a claim for a separate state so much as the secession of a cultural-economic region from an existing state in the context of an integrating Europe. Indeed, it is not always clear that the *precise* boundaries really matter that much in themselves to Bossi or the party militants. The approach seems to be that they are better left somewhat fuzzy, so as to keep one's opponents off guard. After all, in Bossi's world the territory of Padania works outward from the *leghista*'s local base as a phenomenological construct, with the south—sometimes beginning around Rome, sometimes further north, but with Calabria and Sicily as its essential moments—defining what Padania is not. Rather more positively, as reported in a survey of delegates to the February 1997 Congress of the League, when asked how they would define Padania, they replied that it is the "moral values," the "productive system," the "customs," and the "respect for rules" that distinguish Padania from the rest of Italy (Biorcio 1997, 204). The radical "particularism" of Bossi and the League's core supporters betrays itself in the lack of interest in articulating a rational geographical basis—a common past or natural boundaries—for the secession or the increased institutional autonomy that they propose. What is more crucial is the rigid differentiation drawn between a northern approach to politics based on efficiency and transparency and a southern (and Roman) approach based on clientelism and corruption. Territory matters crucially to the political imagination of the League, but from this point of view territory defines a culture associated with a myriad of northern localities rather than an homogeneous Padanian culture defining a Padanian territory. Padania has been invented first, and the cultural strands giving it expression are being selected and interwoven afterward to bring together the disparate norths into one functional unit.

Corresponding to the invention of Padania in 1995–96, the League has also discovered a commitment to Europe as a component of its political project, if not as an end in itself. Roberto Maroni, one of the League's leaders, who was minister of the interior in the Berlusconi coalition government of 1994 and minister of welfare in the post-2000 Berlusconi government, was interviewed by me on 19 September 1997; he sees a connection between the design of Padania and the new attitude toward European integration. He sees Padania as the packaging of everything the League stands for into one word. Padania is a region in Europe which does not aim for the same kind of independence as in the Czech, Yugoslav and Baltic cases. It would be sufficient for Rome to allow autonomy to the northern regions under the umbrella of the EU. Padania is too big for a single identity. It is the objective conditions that the regions within Padania share that is the source of the idea. There is no historical or cultural identity to Padania in the sense that Catalonia or Scotland have. This is our weak point. But if an identity of Padania does not exist then neither does that of Italy. Each is an invention, except that of Padania reflects the needs of northerners for self-government against the power of Rome.

In the creation of Padania, preexisting elements from the past have been put together in a pastiche of deliberate, explicit invention. Bossi has never claimed that he is simply recovering something that once existed but faded away or was destroyed. His is a rather brazen attempt to invent a territorial entity as a vehicle for a set of local identities with little in common save for a common hostility toward the political institutions of contemporary Italy, represented by the single word *Rome*. Critical elements in the invention of Padania include the use of the unfamiliar word *Padania* itself (derived from the Roman name for its imperial province in the Po basin), around which more familiar cultural threads can be woven: reference to the political traditions of the Venetian Republic as a precursor of the present enterprise, reminding potential Padanians of a past in which they were not subjugated by either foreign powers, such as Austria, or to other more Italian forces such as the Popes, Mussolini, and modern governments dominated by "southerners"; and repeated reference to "northern values" of hard work, honesty, and responsibility that distinguish the citizens of Padania from other Italians.

Long before the invocation of Padania, Bossi had broken with the typical language and rhetorical style of the Italian political class. Partly this involved the adoption of a populist vocabulary associated with demonization of politicians and national political institutions, what a leading politician in the old Christian Democratic Party, Mino Martinazzoli, dubbed "the politics of the bar." This appealed to those who saw themselves on the margins of organized politics, alienated from the established parties and angry at the seeming indifference to local demands of the national bureaucracy and its local agents. But it also involved a restructuring of political discourse around a new type of political polarization. Since the Second World War, Italian politics had been structured around such oppositions as left-right, Catholic-lay, workers-bosses, democratic-fascist, and a few others. The League has created a strong sense of *us* for its militants and supporters by breaking with the past and organizing a new political discourse against the *other* of Rome, southerners, and immigrants.

Roberto Biorcio (1997, 204) identifies two dimensions of this discourse. One distinguishes populations and their territories, basically the north and the south, on the basis of very strong normative distinctions between the two. In its strongest form, this opposition is that of Europe (the north) versus Africa (the south). But it also includes the ordinary folk (of the north or Padania) versus the southerners and non-European foreigners. Along the other dimension is an opposition between the high and the low in a hierarchy of social groups; the southern elites are identified with the power of Rome, the old political parties of government, and the mafia, but northern big business, finance, and media conglomerates are also seen as conniving with Roman political power. In Bossi's rhetoric, the media baron and leader of Forza Italia, Silvio Berlusconi, notwithstanding Bossi's on-again, off-again coalition with him, became the main symbol of the system, the oppressor against whom the League must mobilize the people of Padania. Thus, the League combines a set of classic populist themes about the small person exploited by the system with a strong sense of the homogeneously small-fry population of Padania embattled by entrenched interests within Italian political institutions. Padania, therefore, becomes a territorial means of "ignoring the religious fracture and trying to overcome in a populist key the fractures of class between workers, small businessmen, artisans and shopkeepers so as to rediscover unity against big business and a colonizing state which dissipates the resources of northern regions in favor of the South" (208).

Finally, the leader of the party, Umberto Bossi, has been the key proponent of the secessionist strategy, in the face of considerable initial (and in some cases, continuing) skepticism from other leading figures. The more sober and conventionally liberal of them often seem disturbed by Bossi's verbal extravagance, his use of sexual innuendo about his opponents, and his deliberate provocations, including declarations of independence and the creation of parallel institutions for Padania such as a parliament and a national guard. But it is Bossi's style that propels the party into the daily newspapers and has led to public debate among his opponents about the prospects for secession. He even welcomes prosecution as a mechanism not only for publicity but also for "martyrdom" ("La Padania" 1998). His skinny, ill-kempt physical appearance and outrageous rhetoric convey an image of victimization by the powers that be that serves to reinforce the idea of a northern Italy similarly victimized (Barraclough 1998). Padania has provided a powerful vehicle for Bossi's creativity, allowing him to bring together in one relatively concrete territorial concept all of his favorite themes concerning the deficiencies of the present Italian state—in other words, a positive counterpoint to the relentless negativity that previously tended to be his stock-in-trade. Even as he has retreated from secession, partly, it seems, in order to respond to internal critics, but also to find allies to prevent damaging changes in the electoral system, such as a shift to a completely majoritarian one, Bossi has maintained a focus on Padania as the alternative to Italy-as-it-is.

Malleable Political Identities

The idea of Padania and its prospective autonomy or independence seems to have grown rapidly in acceptability among both the militants of the League and significant sections of Padania's electorate. In a mere year and a half following the original declaration of independence by Bossi on 15 September 1996, he carried most of his party with him on the secessionist path. Even such a moderate as Raimondo Fassa, the League mayor of the commune of Varese from 1992 to 1997, acknowledged (in an interview of 8 September 1997, conducted by me) that the "secession idea has met with success. An ignored North has discovered its political destiny because other political currents have not had specific responses to the fiscal and other problems of the region. National unification obviously has no real meaning." Although he goes on to add that there is "limited possibility of independence" and "no cultural unity to Padania," neither is there, he adds, "a strong sense of being part of Europe. For example, Bossi is localistic, not European in outlook." We should note that Fassa expressed these views just ten days before he announced that he would not be a candidate for mayor of Varese in the November 1997 election. He had obviously broken with Bossi. His relatively pure libertarianism, allied with a strong sense of Europe, acquired during his term as a member of the European Parliament, was no longer compatible with the new secessionist strategy, even though he could recognize its propaganda success.

Where Fassa seems to underestimate the Padanian strategy is in assuming that cultural identities are strictly inherited from a distant past and that in the absence of such primordial roots, new political identities are invariably inauthentic. This, of course, is the "common sense" of much contemporary social science. What is remarkable in the case of northern Italy today is that we can see before our eyes the attempted invention of an identity that has had no prior existence. What the latest turn in the League's manipulation of territorial symbolism suggests, therefore, is that political identities are much more malleable or subject to revision than most social scientists and historians have tended to think, particularly in the presence of powerful aversions to existing institutions and fear that these institutions no longer defend or further local interests. The difficulty, as a former adviser to Bossi, Gianfranco Miglio, has put it, has been in turning such local interests into an identity that can effectively mobilize people politically (Miglio and Barbera 1997, 168–69). In his detailed examination of opinion polls among League activists and northern voters, Roberto Biorcio (1997, 131) claims that by 1997 "[t]he invention of the nation of Padania, considered by many to be without any foundation whatever, has started to put down roots, if still among a minority, in the electoral sphere of the League, and in general in public opinion in the regions of northern Italy."

How did this happen? It started as recently as the mid-1980s when Bossi and the Lombard League (as Bossi's League then was called) publicly challenged the Italian territorial status quo. A 1984 study of political participation in Lombardy showed that sentiments of local and regional belonging or identity were much less important than other types of identity, based on class, occupation, status, religion, and demographic characteristics such as age, generation, and marital status (Biorcio 1997). Only 18.3 percent of the respondents to a survey reported that any kind of territorial identity (national, regional, or local) was more important than others, whereas over half (53.1%) said that social positions of one sort or another were most important. Among the territorial identifiers, the most widespread attachment was to the national level. The local and regional identifiers were widely seen at the time as exemplifying the "localistic" type in Robert Merton's famous distinction (1957) between locals and cosmopolitans: people without much interest in politics, church-going, and interested in local gossip, with the regionalists somewhat more sophisticated politically than the average localist. In other words, they were the remnants of *Gemeinschaft* or community in a world that was increasingly that of *Gesellschaft* or geographically wider-ranging society.

To track shifts in territorial identities, there is evidence from three Italy-wide surveys of young people between the ages of fifteen and twenty-nine conducted in 1986, 1991, and 1997 and three general population surveys in northern Italy conducted in 1992, 1994, and 1996. From these surveys it is not possible to weigh territorial against social identities but only to see to what extent there have been changes in the nature of territorial attachments. In 1986 throughout Italy there was a tendency on the part of young people to privilege local attachments. Regional attachments were much lower. Somewhat ironic in the context of later changes, northern respondents were more likely to report an identity with Italy than were those in the center and the south. By 1991 local identities had weakened considerably, with a commensurate rise in the importance of all others (national, Europe, the world, region), with region having the largest increase in the north, going from 11.4 percent in 1986 to 15.1 percent in 1991. This trend was particularly noticeable among young people voting or intending to vote for the League. Close to one-third of them reported a priority attachment to region; a similar number expressed local attachments, and around 24 percent expressed an attachment to Italy.

In a 1992 survey of northern residents, fully 46.1 percent favored large-scale fiscal decentralization for the northern administrative regions, with shopkeepers, artisans, and workers giving it the strongest support. For these social groups, a regional identity seemed to increasingly capture their social identities. This trend deepened in two more recent surveys in 1994 and 1996, in which a net majority of all residents expressed support for northern autonomy, far more than actually vote for the League. Although local attachments remain the most common, realizing these is seen as requiring regional autonomy. Among League supporters, however, the territorial sphere to which they feel most linked has shifted from the local to northern Italy as a whole. In 1996, 46.9 percent of League sympathizers reported a primary identification with northern Italy, compared to 47.6 percent identifying with their local commune. Indeed, close to a quarter of the voters for the two main right-wing parties Forza Italia and Alleanza Nazionale also expressed this preference. A sizable proportion of the northern population, around 50 percent, expressed either a first or second choice for a primary northern Italian or northern regional identity (Lombard, Venetian, etc.). This comes very close to challenging the dominance of local identities and is in excess of the primary or secondary preference for an Italian political identity (44.1%).

The growing sense of attachment to northern Italy or Padania is reflected also in the changing levels of support for different institutional arrangements. Around 11 percent of the residents of northern Italy are in favor of a strong autonomy for Padania as part of a confederation or as a macroregion in a federal state. The same opinions were expressed by 39.8 percent of the League's sympathizers and 29 percent of its voters. Of all the respondents, 36.4 percent would like an Italian federation of administrative regions (34.6% of the League's sympathizers and 41.8% of its voters). Very few League supporters would favor enhanced powers for local entities, whereas among other political groupings such a move tends to be much more popular. This conclusion is reinforced by comparing the findings of a survey of participants in the March to the Po in September 1996 with that of delegates to the League's Congress of February 1997. Not only did sentiments in favor of independence increase from 65.9 to 74.9 percent, but the strongest sentiment of attachment was even more strongly for Padania than for the other levels offered, such as city, region, or Italy. There is a danger, however, in projecting such findings from activists onto the population of northern Italy at large. The majority of poll respondents in northern Italy neither want nor expect secession. At the same time, those respondents who have the strongest identification with the north or Padania do see secession as a threat that can lead to the empowering of regional and local levels of government (Biorcio 1997, 121–25, 202–3).

At the very least, therefore, the proposal of Padania has pushed the general population and the other political parties toward favoring more powers for local government. For example, only 27.4 percent of respondents to the 1996 survey expressed support for the present system of government. Among many League supporters and an increasing number of other northerners without strong affiliations to the League, the substitution of the north for such administrative regions as Lombardy and the Veneto and the invention of Padania seem to have produced a veritable shift in dominant political identity from the local area to that of Padania. It remains to be seen, however, whether such sentiments will ever translate into the kind of vote that will be required for secession to achieve success, given that they seem largely confined to localities within the north that are already both strongly localistic and pro-League. Bossi's equivocation about secession beginning in August 1998 and the defection of many of the

League's local leaders in the Veneto to the Liga Veneta suggest that such a choice may never have to be made.

Multiplicity of Identities

It is misleading, however, to give the impression that a new Padanian identity is simply replacing older ones in a type of parade or serial progression of political identities. In fact, each of the surveys I have referred to presents evidence that people have relatively complex political identities, in which a number of territorial and social dimensions intersect. There are two aspects of the Padanian case that are worth identifying. The first is the ease with which people maintain a number of political identities even as they shift them in order of priority. In contemporary northern Italy (and Italy as a whole), local and national identities have long coexisted without any sense of mutual exclusivity. Each of the surveys of Italian youth, from 1986 through 1991 to the latest in 1997, demonstrates this. When asked to give in order of importance their two most important geographical units, not only a hierarchy of levels but also combinations of attachments always arise. Although the local appears consistently in these surveys as the most important territorial identity everywhere in Italy, two-thirds of those who indicate that preference also consider Italy as a complement rather than a competitor to it. In the words of Ilvo Diamanti (1997, 49), one of the most careful students of the League and its impact on Italian identities, this mixing of territorial identities is "[a] sort of frame which permits a country of localisms and localists like Italy to stay together, even if with conflicts and particularisms."

In such a core area of the League as the northeast, for example, in 1997 attachment to Italy did not register support above the national average as a primary affiliation, but it did have a higher affiliation than any major regional division of Italy as a secondary attachment. Indeed, it is among higher-status social groups and supporters of left-wing parties that there is the least pride in being Italian and higher commitments to the world or Europe. Most young Italians, including supporters of the League, manage to combine a high degree of municipalism or localism and a limited regionalism with a significant sentiment of national pride (see table 9.1). A preference for the League and for Padania can readily coexist with a significant, if hardly dominant, sense of national identity and others such as European identity. At least for younger Italians, therefore, the local and the cosmopolitan are not necessarily in opposition, as conventional sociological thinking might lead us to expect. Rather, "Italy continues to suggest, as Paolo Segatti has proposed, a nation of *compaesani,* or fellow local dwellers, who look to Europe and the world with great attention, but with limited passion" (quoted in Diamanti 1997, 53). Passion is now far closer to home.

The second aspect of the Padanian case in relation to a multiplicity of political-territorial identities is that the League's leaders appear conflicted about its primary territorial orientation. They seem to share the localism of the population at large. The League's very success in local elections in northern Italy also now means that the League is institutionalizing itself. League politicians now dominate many communes and provinces. They tend to believe that the electoral future of the League, and their own retention of office, lies in delivering good government at the local level or building a new northern Italy "from below." This inevitably seems to privilege the existing local government units at the expense of the vision of a northern macroregion such as Padania. But, rather like the Communists in central Italy in the 1970s, when exclusion from national governments led them to invest heavily in creating efficient and honest local government so as to advertise their qualifications for national office, the League at the local level can use its control over local administrations to tout the possibilities for government throughout the north when the evils of Roman rule are finally removed. This is the point of view encountered most frequently in formal interviews that I conducted with local politicians in Varese, in both September and December 1997. Massimo Ferrario, the president of the province of Varese from 1995 until 2002 (interviewed on 11 September 1997), for example, is an important League politician who sees the local and the macroregional as complementary: "The institutionalization of the League in local agencies," he says, "allows for an active growth of federalism because it is impossible that the Italian state will reform itself to allow for a true federalism. Provincial autonomy is a goal shared by all presidents of provinces but it can only happen under the threat of leaving Italy behind." In other words, Padania and increased autonomy for local institutions are seen as going hand in hand. The local and the regional, therefore, are complementary more than competing. This can lead to a technocratic or good-government ethos that actively undermines the confrontational strategy of Bossi. Such an ethos could develop through an absence of zeal for Bossi's latest example of political "performance art." It is more likely

TABLE 9.1 Italian National Survey of 15–24-Year-Olds, December 1997, Percent Responses

	First Choice	Second Choice	Total
The city in which you live	40.2	19.2	59.4
The region or province	10.5	23.2	33.7
Italy	32.2	31.8	64.0
Europe	3.1	13.1	16.2
The world in general	12.6	10.6	23.2
No reply	1.4	2.2	3.6
	100.0	100.0	

SOURCE: *Luce* 7 December 1997, 4.

NOTE: The question asked was, "To which of the following units do you have the strongest attachment? And, in second place?" Totals of first and second choices do not add up to 100 percent.

to develop, however, through the need to build local political coalitions. In the commune of Varese in the period 1992–97, for example, the shortage of qualified League representatives to serve as administrators led to the appointment of former Communists and others in an "anomalous" municipal government that aroused much comment and not a little ridicule among the League's opponents (Pajetta 1994).

Conclusion

Understandings of the connection between territory and political identity have suffered from assuming that state-scale territoriality is primary in processes of identity formation. This chapter began by tracing this orientation to two important but largely hidden assumptions of modern social theory: (1) the moral identity assumed between statehood and personhood and (2) the conception of identity-formation as involving pregiven entities, individuated persons, who are then socialized at the level of the state into national identities. After questioning both of these on theoretical and empirical grounds, I traced the relationships between different territorial scales of political identity in contemporary northern Italy.

My purpose here is to show that the connection between territoriality and political identity is much more historically contingent than conventional accounts would have us believe and involves a number of different processes of identity formation. Three such processes are detailed for northern Italy: the invention of a new regional level of identity, the malleability of political-identity formation, and the coexistence of multiple political identities. Much more complex understandings of territoriality as involving different scales of social life and the balance between them are necessary if the sociology of political identities in contemporary Europe is to be adequately understood.

NOTES

1. For example, see Trubowitz's interesting case (1998) for persisting regional sectionalism in American foreign-policy making.

2. See, for example, Colley 1992 and Nairn 2000. More generally, see A. D. Smith 1991.

3. This is also the main point of Brubaker and Cooper 2000. It leads the authors to want to abandon the word *identity* altogether because in much usage they see it as avoiding the processes whereby identities are defined in favor of fixed and essentialized categorizations.

4. Anderson (1991), the inventor of this term, makes no reference to the territorial element in the construction of political identities.

TEN

The Democratic Integration of Europe

Interests, Identity, and the Public Sphere

CRAIG CALHOUN

The term *public sphere* is a spatial metaphor for a largely nonspatial phenomenon. To be sure, public spaces from the Greek agora to early modern marketplaces, theaters, and parliaments all give support and setting to public life. But public events also transform spaces normally claimed for private transactions—as parades transform streets. The public sphere is a "space" of communication, and as such it transcends any particular place and weaves together conversations from many. Publics grow less place-based as communications media proliferate, yet the spatial image remains apt.

As Hannah Arendt wrote, public speech creates a space among speakers and the possibility of institutional arrangements that endure beyond the lives and mere quotidian interests of those speakers (Arendt 1958). If Europe is not merely a place but a space in which distinctively European relations are forged and European visions of the future enacted, then it depends on communication in public, as much as on distinctively European culture, or political institutions, or economy, or social networks.

Public communication takes place in, and helps to create, a space of relationships among citizens. This space is not the whole of relatedness; it is only one domain or realm among many. And as Jürgen Habermas (1989) noted, *realm* and *domain* (which have their own spatial connotations) are in one sense precisely wrong, because the public sphere is constituted by the multidirectional communications of its participants, not any rule from above.

Such communication—the public sphere—has at least three dimensions that are important for European integration. First, it enables participation in collective choice, whether about specific policy issues or basic institutions. Second, public communication allows for the production, reproduction, or transformation of a *social imaginary* that gives cultural form to integration, making Europe real and giving it shape by imagining it in specific ways. Third, the public sphere is itself a medium of social integration and a form of social solidarity, as well as an arena for debating others.

The self-constitution of Europe through public communication is a relatively neglected aspect of European integration. Yet, because it brings a unique and crucial condition for collective choice, it is basic to the possibility that Europe's integrated institutions can be democratic. The public sphere is also an important counterpart to forms of integration based on struggles and negotiations among specific interests. First, it opens the possibility that actors may redefine their interests in the course of public communication and shifting understanding of both collective good and individual or collective identity. Second, it offers the potential of a space constituted not as the sum of particular territorial locations, or political-economic locations, but by communication among strangers, addressing the public as such.[1] The growth of the public sphere in early modern cities was thus a different dimension of civil society from the growth of guild organizations or marketplaces, though related to each. It constituted an important aspect of the city itself as distinct from a location within it. Similarly, within the processes of national integration, public spheres played crucial roles in constituting the nation as such, as object and arena of discourse, distinct from the particular interests, regions, face-to-face communities, and nodes of activity within it. How much this will happen on the scale of Europe remains to be seen.

The questions are not all about the relative organizational capacities of the EU and its member states. They are also about the relationship of both to transnational processes not contained within Europe (which we may call global without implying that they obtain equally all over the world). They are also about the extent to which public life thrives at the geographical level of cities and the ways it is (or sometimes is not) produced by means of space-transcending media. Though it may be physically impossible to be in two places at once, it is not impossible to inhabit simultaneously several of these metatopical spaces of public communication.

The Problem of Integration

Academic debates over European integration engage an impressive range of scholars. Some question whether there is societal integration to match economic or administrative integration and accordingly raise questions about legitimacy. Many question whether European integration necessarily comes at the expense of national identities and, if so, whether these are really fading. Others question what constitutional form might be created not only to legitimate but to govern the emerging polity, to shape how it will balance democracy with technical administration and judicial review. Still others take up the substantial extent to which European integration has advantaged capital, which moves freely, and disadvantaged labor and other social groups that remain both fragmented and bound by national laws.

In this chapter I touch on these issues but address more centrally the questions of what role participation in the public sphere is playing in European integration, what role it could play, and how much actual patterns of integration are furthering the development and effectiveness of a European public sphere. I am optimistic on potential but pessimistic on actual achievement (and my emphasis is more on theory and conceptualization than on empirical measurement). I also try to distinguish participation in the public sphere as one form or modality of integration from others—such as integration of markets, or functional integration more generally; development of a common culture or identity; creation of wider-reaching and more diverse networks of interpersonal relationships; and the sheer exercise of power, notably by states and the new EU administrative apparatus, but also by some economic actors.

For the most part, I do not enter debates concerning the overall pace of European integration. Such integration is happening, even though how fast, through what institutional forms, and to whose benefit can all be debated and are subject to further research. Integration could yet be diluted by the addition of new member states, and the legal framework of integration continues both to be disputed and to evolve. To what extent a more united Europe will remain organized by treaties among sovereign states or will itself become an integral state remains uncertain. A variety of creative prospects for shared sovereignty excite political theorists and may open a space in between creation of a new state and mere agreement among older ones. At the same time, it is clear that the EU's member states spend vastly more of the region's gross domestic product, wield more military force, and play larger roles in most governmental affairs, save regulation of interstate commerce. While the EU has not yet become a very effective foreign-policy actor, there is an important European voice in global affairs. This is articulated by leaders of European states, individually and interactively with each other. It is also articulated in nonstate public forums by social movement activists, newspaper editorialists, academics, and other participants in public discourse. It is recognizable and recognized, even though it is not unitary in either content or institutional base.

Affirming that integration is real and substantial does not answer the question of just how Europe is—or is to be—integrated, nor how the specific ways in which Europeans are joined to each other affect the prospects for democracy, collective choice about the future, and recognition of difference within unity. Integration is not a simple good, always equally desirable whatever its form. Rather, the case of Europe offers a good empirical focus for considering the implications of different forms of social integration for democratic politics.

European integration has been driven importantly by two negative projects: avoiding war and avoiding American (and sometimes Asian) economic hegemony. These have been justified mainly by appeal to presumably widespread interests. They (and the building of European institutions) have been implemented largely on the basis of agreements among the governments of member states. Insofar as the member states are democratic, this process is not intrinsically contrary to democracy. The actual operations of the new European institutions, however, have often turned out to be less transparent, less accountable, and less amenable to popular participation than those of most member states. They reflect the influence of an internal technocratic elite, and they also appear to be more susceptible to the influence of business corporations and closely related external elites. This is why they are criticized for a "democratic deficit." At the same time, European integration has been a source of new inequalities between business and labor and between a cosmopolitan elite and less mobile (often less transportably credentialed) citizens. Professionals, including academics but also lawyers, accountants, and others, have been among the beneficiaries of European integration. Even where they have not been among its leading advocates (as they often have), they have usually adapted fairly readily to it and have found new opportunities within or in dealing with European institutions. Skilled workers, by contrast, have often found less transnational recognition of their credentials and fewer opportunities. And while European integration has opened local markets to transnational corporations, labor law has remained mostly national and thus fragmented (Streeck 2001). Workers who move generally derive less benefit and protection from the EU than do firms that move capital. For all these reasons, the EU suffers also from a deficit in its perceived legitimacy. One indicator of this is the fairly consistent Eurobarometer finding that shows much greater support for "European integration" than for "membership of the EU."

It is partly for this reason that appeals to a common European identity have increased to complement longer-standing appeals to common interests. While these commonly appear as transnationally cosmopolitan within Europe, they are also often linked to resistance to non-European immigrants (though of course, ***European*** is a contested identity in this regard, e.g., vis-à-vis eastern Europeans and Turks). This is a source of what might be called Pym Fortuyn's paradox—the claim that in order to protect liberal society, one must illiberally resist immigration. While political specifics vary among nations, ideologues, and movements, the paradox itself features in most European countries and at the level of Europe as a whole (and to some extent in most of the world's more or less liberal societies). In any case, if internal coherence and external closure are basic features of the claim to be a society, then appeals to common identity do double work by helping to bolster the legitimacy of the internal institutions that produce coherence while also giving an account of why borders should be closed to immigrants (though not, despite occasional populist gestures, to capital).

Appeals to European identity grow still more important to the extent that European institutions affect citizens in direct, not merely indirect ways. For most of the first forty years of postwar integration, the effects were mainly indirect. The EU administered an increasingly expensive common agricultural policy, but the effects were filtered through member states and national prices. It created regulations that were administered by member states. Especially from the 1990s, though, the EU and European integration became more manifest in everyday life. EU citizenship changed queuing in airports, for example, and the euro replaced national currencies. In foreign assistance and some other parts of foreign affairs, the EU struggled to present itself as a unitary and autonomous actor. While the range of "Euro-goodies" (including academic grants and exchanges) flowed most to elites, ordinary citizens were also now called upon to think of themselves as European (Schmitter 2001). There is debate about how much they do so, underwritten by fluctuating survey data. The fluctuations themselves are not surprising; there is no reason to expect European identity not to be in considerable part situational (Hedetoft 1997). After all, this is true of most identities. National identity is commonly understood as being immutable, but in fact its salience varies—increasing with travel, confrontation with immigrants, and war—and its character is neither fixed nor the same for all nationals.

Regardless of how readily European citizens self-identify as European, there are also a variety of ways in which behavior is becoming increasingly similar throughout Europe. Slang, clothing styles, music, and movies all circulate more widely. This may lubricate European integration. Mere similarity, though, is different from mutual interdependence (as Durkheim [1893] 1984 famously argued) and is a very segmentable form of solidarity. Before the nineteenth-century heyday of national integration, the internal regions of European countries were often more diverse than the whole countries were from one another—on dimensions ranging from wealth to birthrates and family formation (Watkins 1991; Weber 1976). On some dimensions of similarity and difference, this may be happening again. At the same time, many similarities among Europeans reflect broader cultural and market flows and common circumstances. This is not only a matter of consumer culture but also of learning English. Thus, some of the ways in which Europeans are most similar, and are growing more similar, are not specifically European.

In short, there are many limits to common identity as the basis for either legitimacy or solidarity in Europe. European identity could grow stronger and more uniform without supporting democracy. It could constitute nationalism on a continental scale. Common identity is, in any case, only one of several forms of social integration; others include markets and other autopoetic systems, networks of interpersonal relationships, and domination by those with one or another form of power.

It is common to analyze the extent of European integration—including questions of whether Europe is becoming a single society or polity—in terms of internal coherence and external closure. Both are important; pointing to the persistence of national difference or the importance of global flows challenges each. It is equally crucial to account, however, not only for coherence and closure (both of which can of course be no more than relative) but also for mutual commitments among the members of the polity—including commitments to the fairness of political processes. This is a central point of Habermas's appeal to constitutional patriotism.

Constitutional patriotism depends on a vibrant public sphere. As Habermas (1998, 160) says, "From a normative perspective there can be no European federal state worthy of the title of a European democracy unless a European-wide, integrated public sphere develops in the ambit of a common political culture." In this public sphere, citizens will join in debate over the kinds of social institutions they want. But we should not think of the public sphere as only an arena of rational-critical debate or of common political culture as formed in advance of participation in the public sphere. Rather, culture—and identities—will be made and remade in public life. Building on this, I wish to urge a richer conception of the public sphere as an arena of cultural creativity and reproduction in which society is imagined and thereby made real and shaped by the ways in which it is understood. It is because public life can help to constitute a thicker, more meaningful and motivational solidarity that it can help to underpin a modern democratic polity. A thin identification with formal processes will not do.[2] Citizens need to be motivated by solidarity, not merely included by law.

The problem of European integration, then, is not simply to achieve solidarity or to make an effective union. It is to do this in a way that is conducive to democracy and fairness—and other normative values that citizens might choose or develop. Not all integration is equally benign; not all that is benign in itself is helpful for democracy.

■ Necessity and Choice

To what extent can the continuing formation of European society be based on widespread, popular, democratic choice? I say *formation* advisedly, to emphasize the greater stakes in a process commonly discussed under the milder label *integration*. There is no such thing as a neutral integration, in which the various countries, regions, cultures, peoples, economic systems, and social movements of Europe fit together as though they were pieces of a jigsaw puzzle. Each is unquestionably changed by integration with the others. Moreover, the social whole being created is not simply a sum of its parts, however integrated, but something new. This new entity is the product of several different processes of integration and creation. Its own overall form is shaped by a combination of conscious choice, a less conscious social imaginary that influences actors' sense of what forms are possible, and various material and symbolic processes in which actors engage for other reasons but with more or less unintended consequences for the formation of European society. The latter include engaging in markets; broadening consumer tastes; building transportation and communications infrastructures; attempting political domination; producing, circulating, and acquiring cultural goods; and participating in social movements.

Choice has not been unimportant so far in the formation of Europe, but (1) the choices of elites have been vastly more important than those of broad populations, (2) the self-conscious choices of broad populations have been mostly limited to yes/no referenda rather than choices of form, and (3) the most influential choices of broad populations are not self-consciously about the formation of European society but about a range of everyday activities in which they respond to particular interests and values in ways that shape Europe. These last range from buying food, listening to pop music, and cheering football teams to demanding pay raises, complaining about bureaucrats, and responding to immigrants with acceptance or hostility. Getting people to use euros proved easy; they integrated readily into everyday life. Getting people to care about elections to the European Parliament has been harder; the results have seemed remote from most people's lives (whether because they rightly analyzed the limits on parliamentary power within the EU, or because they more questionably judged the EU itself to be a minor influence, or because they thought simply that they could have little influence by voting for the actual candidates presented to them).

Moreover, much discussion of the reasons for European integration stresses not choice but necessity. The discussion not only presents integration as necessary; it presents specific institutional forms of such integration as either necessary or at least to be assessed on grounds of technical efficiency rather than explored in terms of social consequences. The necessity of, say, resisting the "American model" (any American model, in media or markets or multiculturalism) is presented less as an occasion for choice among the potentially innumerable other models than as justification for specific institutional arrangements and policies. Yet institutions are sustained and shaped not only by mechanisms of reproduction that individuals cannot alter, but also by choices of participation, rejection, and struggle.[3] Of course, many choices are highly undemocratic, reflecting differences of power. Nevertheless, many, perhaps especially in modern European countries, are potentially open to more or less democratic processes. The rhetoric of necessity obscures this. Once centered on preventing war, the rhetoric of necessity is today overwhelmingly economistic. However, as Larry Siedentop puts it, "If the language in which the European Union identifies and creates itself becomes overwhelmingly economic, then the prospects for self-government in Europe are grim indeed" (2000, 32).

I propose first to distinguish choice from mere response to interests or necessity, on the one hand, or reflection of collective identity, on the other. Interests and identities certainly influence choices, but I want to defend the idea that choices are actions (individual and collective) that imply the availability of multiple possibilities. The choice itself therefore matters and is not strictly determined by preexisting conditions. Choices are shaped by social imaginaries—that is, more or less coherent sociocultural processes that shape actors' understandings of what is possible, what is real, and how to understand each. The influence of both interests and identity is refracted through such imaginaries—thus, not simply through culture generally but through specific formations that naturalize and give primacy to such ideas as ***individual, nation,*** and ***market.***[4]

Second, I want to suggest the central importance of the public sphere not only as an arena in which individuals debate collective choices, but as a setting for communication and participation in collective action that can shape identities and interests, not only reflect them. Many treatments of the public sphere, including Habermas's classic *Structural Transformation of the Public Sphere* (and also much of the early modern political theory on which it was based), have approached it as an arena in which individuals fully formed in private may communicate about public affairs. These individuals have "private" identities and interests, which they ideally set aside in order to maintain high standards of rational critical discourse. On such an account, the public sphere is not an arena in which culture and identities are formed and in which participants may redefine who they are and what interests move them. Indeed, if it takes on this culture-forming character, Habermas (1989) sees this as a problematic dedifferentiation or falling away from its more rational potential (compare Calhoun 1992 [ed.], 1995; Koselleck 1988; and Warner 2002). The focus is on the sociocultural bases for sound collective judgment. Habermas's model has the considerable advantage of offering an account of how actual social inequality might be kept from distorting public discourse (by disqualifying diversity of interests and identities). A disadvantage is that this does not seem to have worked very well (his book is about the twentieth-century "degeneration" of the public sphere as well as the eighteenth-century ideal for it). Moreover, the observable formation and reformation of culture, selves, and interests in public life means that these are themselves potentially open to choice.[5] The women's movement offers ready examples, but, though less commonly remarked, the same process is also important in the shaping and reshaping of ethnicity. With regard to Europe, this means that mutual engagement in public life need not depend on prior cultural similarity or compatibility of interests.

Third, but in very close relation, I would argue that participation in public life can accordingly be itself a form of social integration or solidarity. I do not mean that this is necessarily the outcome of all public communication, or that such communication is always harmonious. I do mean that participation in the public sphere integrates people into discourses and projects and collective understandings that connect them to each other. It is literally *voice,* in Hirschman's terms, but voice that can engender loyalty—if not always to what is created, at least to the process of collective participation. When the public sphere is active and effective in shaping the choice of social institutions, it invites even those who would reject or change existing institutions into a collective process of mutual engagement with others—a form of solidarity whether or not it is precisely one of harmony. The argument I present is thus related to Habermas's notion of constitutional patriotism, but I want to stress the importance not only of loyalty to created institutions but of participation in the process of creation and recreation. So far actual European integration has relied less on such a public sphere than might be hoped and more on other ways of achieving connections among people, organizations, and states. These other ways—such as functional integration, sheer exercise of power, development of a common culture, and formation of more diverse interpersonal relationships—are not necessarily bad but do not offer the possibilities for democratic choice inherent in a public sphere.

■ Interests and Identities

A variety of prominent and influential actors have clearly chosen European integration. States have signed treaties, and statesmen have issued ringing declarations. Some, like Jacques Delors, have committed their careers to it. Others, like Helmut Kohl, have invested their political fortunes in it. Even among elites, however, choice has often been disguised in a rhetoric of necessity: the necessity of avoiding war, of choosing sides in the cold war, or of competing in global capitalism.

Necessity is an extreme form of argument based on interests, in which interests are understood to be both associated with specific actors and their social circumstances and objectively constraining or determining. Thus it has been argued — successfully — that Europeans, or European countries, or European corporations, have an interest in more effective integration. Having established an interest in integration, there can then be arguments about what form of integration better serves the interests of actors (noting the potential tension among different kinds of actors with different amounts of power and influence). Should the integration be federal or confederal? Should it give more power to the European Parliament, or the Commission, or the Council of Ministers? Should it include more integration of labor laws or corporate laws, or of markets or legal sovereignty?[6] So long as this is understood as an argument about serving interests, it can be pursued in a more or less utilitarian rhetoric. The best solution is that which maximizes aggregate interests. There is, however, a catch that arises from the differences among recognized actors: persons, corporations, and states not only have different degrees of power with regard to pursuing their interests, but their diversity of form creates a problem in aggregating interests according to the Benthamite maxim of seeking the greatest good of the greatest number. By what calculus does one integrate into a single equation the interests of a corporation and an individual person? They would seem to be incommensurable.

This problem is solved, at least in academic treatises and some political and economic rhetoric, by asserting that ultimately only individuals have real interests. It was in this sense that Margaret Thatcher asserted that there is "no such thing as society." Thatcher echoed Jeremy Bentham, who wrote that "the community is a fictitious body composed of the individual persons who are considered as constituting as it were its members. The interest of the community then is, what? — the sum of the interests of the several members who compose it" (Bentham 1982, 12; see also Mansbridge 1998; Calhoun 1998b). Conceivably, Thatcher and those who follow her line of reasoning would argue equally that there is no such thing as a corporation or a state — but only the individuals that make it up.

This is, in a sense, the point of contractarian theories in which both corporations and states are presumed to be created simply by the voluntary choices of autonomous individuals entering into contractual relations with each other (and states are thus treated as simply a special form of corporation). Simply posing the question reveals that arguments from interests depend also on arguments about identity. The latter establish the subjects that have and act upon the ostensible interests (and thus implicitly, the interests themselves—as Coleman 1989 noted, though one need not follow him in arguing that identity itself can be analyzed internally to rational choice theory).

Individualism has certainly been the most widespread rhetoric of identity, constituting egocentric persons as putatively autonomous actors, with interests of their own. Almost as prominent as individualism, however, is nationalism. Margaret Thatcher was not above appeals to British patriotism (say, in the context of the Falklands-Malvinas War) that went beyond treating the country as a merely contractual and voluntary arrangement among autonomous individuals. More generally, nationalist assertions of collective interests have been vital both to securing the compliance and support of non-elites and also to motivating elites themselves. These establish nations or peoples as unitary subjects (individuals of a kind), which have interests, are represented by states, and grow richer or poorer in international economic accounting (e.g., of GNP, GDP, or various reckonings of development and life chances). Following this rhetoric of identity, one may assert the primacy of nations as contracting partners in European institutions, while denying autonomous existence to Europe (in the same way that the Thatcher-Bentham position denied that communities or societies were anything other than the sum of their members). Or, however, one may also assert that Europe itself has or could have the status of a unitary whole, as an encompassing polity not reducible to the sum of its members. Such arguments fuel the considerable industry of debate and research on questions of whether there is, or is coming to be, a European identity. For modern ideas of legitimacy depend on the notion that a government or political power serves the interests of its people (whether we understand these strictly as legal citizens or in some more encompassing way; see Calhoun 1997; Taylor 2002). Either directly at the level of Europe, or indirectly at the level of constituent states, thus, there must be an appeal to some determinate link of people to state. This is what nationalism commonly provides—not only in legal criteria for membership but in accounts of a common history, participation in a community of fate, or dependence on a shared culture.

To speak of an interest in European integration, thus, is to enter a discussion in which establishing the identity (or identities) of relevant actors is crucial. At the very least, the two are coeval and mutually interdependent. For particular analyses, it may make sense to treat one as underpinning the other: identification of actors with each other may be a source of shared interests, or common interests may lead to a greater sense of shared identity. There is, however, no escape from the need to establish identity at some level in order to establish interests. For interests to be anything other than completely ephemeral preferences, they must be analyzable in terms of the identities of social actors — interests as a woman, for example, or as a worker, or as a Muslim. But while there is an aspect of actors' identities that is a more or less objective reflection of social position and shaped by processes of ascription, there is also a politics of identity and a personal process through which people do or do not take up various possible identities and thus give them differing meaning and variable salience.

Identities (as well as interests) are influenced by material social processes. For example, the nineteenth-century development of transport and communications systems substantially increased the density of linkages within countries. It contributed to an increase in relatively long-distance trade within state borders. This very likely helped to foster a shared sense of national identity. So did military service, national education systems, and the development of newspapers with national readerships. All of these helped to produce not only more "sense" of national identity but greater actual similarity of behavior within nations and difference from neighbors. As Watkins (1991) shows, for example, fertility behavior (including not just birthrates but legitimacy and age of first birth) varied more among provinces within countries up to the early nineteenth century than it did in aggregate between countries. By the mid–twentieth century, this pattern had been reversed. The change was due apparently to sharing of both knowledge and norms within communicative networks that were increasingly organized on a national basis. Conceivably, new media and other networks of European integration could reverse the pattern again, reducing national differences while allowing regional and local ones to grow.[7]

Identity is a matter of both actual similarity and self-understanding. One can influence the other, and both can change. The idea that national identity, for example, is simply an ancient inheritance is clearly false. It was itself made and continues to be remade in sociocultural processes that could in varying degree be paralleled on a European scale. Bits of this are already apparent—a progressive rewriting of national history texts to include more of a pan-European story, growing exchange of students among European universities, linked editorials in newspapers, and shared entertainment television (though not, so far, a great deal of it).

Some writers privilege the processes by which nations have already been formed, suggesting that it is intrinsically impossible to duplicate them on a larger scale, for example by means of new media of communication. This seems not merely to underestimate new media but to exaggerate the ways in which older national identities were formed partly in technologically mediated communication and indeed formed on very different scales, some larger than the EU. According to Anthony Smith, "no electronic technology of communications and its virtual creations could answer to the emotional needs of the 'global citizens' of the future, or instruct them in the art of coping with the joys, burdens and pain and loss that life brings" (2001, 136). But this is comparable to saying that speaking or reading could not provide the instruction. It confuses medium with content and social context. Electronic technology surely could help mediate emotional attachments to large categorical identities—nations, for example, or religions—and carry messages that answer to emotional needs and offer instruction. It is unlikely that it could be effective in this standing alone, without the complement of face-to-face interaction, any more than newspapers have been. That is a different matter, though, and in itself does not speak to the scale of attachments or identities.

Some sort of appeal to preestablished identity is implied by the way even Habermas (1998, 141), in common with much of political and legal theory, poses this basic, orienting question: "When does a collection of persons constitute an entity—'a people'—entitled to govern itself democratically?" The suggestion of a temporal order, that the people must somehow mature into readiness for democracy, has been a staple of dictators and elites who think popular democracy should be postponed to some future in which the people are ready. It is importantly challenged by the suggestion that democratic participation is itself the educative process that potentially makes a people ever more ready, not only by developing democratic capacities, but also by developing democratic solidarities. In this regard, Habermas is rightly critical of the kind of answer to his orienting question that is typically incorporated into the idea of nation-states:

> In the real world, who in each instance acquires the power to define the disputed borders of a state is settled by historical contingencies, usually by the quasi-natural outcome of violent conflicts, wars, and civil wars. Whereas republicanism reinforces our awareness of the contingency of these borders, this contingency can be dispelled by appeal to the idea of a grown nation that imbues the borders with the aura of imitated substantiality and legitimates them through fictitious links with the past. Nationalism bridges the normative gap by appealing to a so-called right of national self-determination. (141)

Nationalism disguises the extent to which both borders and solidarities are creatures of power. To place nationalism on the side of "mere history," and thus implicitly of power without justification, is to encourage too thin a view of culture. To see civil society as simply a realm of voluntary action is to neglect the centrality of systemic economic organization to it—and of the public sphere to the self-constituting capacity of civil society. To see the public sphere entirely as a realm of rational-critical discourse is to lose sight of the importance of forming culture in public life, and of the production and reworking of a common social imaginary. Not least of all, both collective identity and collective discourse depend on social organization and capacities for action—whether provided by states or civil society.

There is no intrinsic reason why either nationalism, even nationalism with strong ethno-cultural components, or constitutional patriotism could not flourish on the scale of Europe. Both states and markets are already contributing to the production of European identity, as both contributed to the development of national self-understandings earlier. The EU does engage in some of the kinds of knowledge-producing practices that have previously been used by states to render regional diversity an aspect of national unity: producing maps, surveys, inventories and celebrations of heritage, administrative classifications, and so forth. It has not yet gotten into the museum business in a major way, but museums not run by the EU itself have organized exhibitions that present Europeanness. The capacity of the EU to guide such processes remains low, however, relative to its constituent states, in a way that was generally not true for earlier national states relative to constituent provinces (though obviously these had more power in some settings—notably Germany—than others).

Habermas (1998, 115) suggests that "the nation-state owes its historical success to the fact that it substituted relations of solidarity between the citizens for the disintegrating corporative ties of early modern society. But this republican achievement is endangered when, conversely, the integrative force of the nation of citizens is traced back to the prepolitical fact of a quasi-natural people, that is, to something independent of and prior to the political opinion- and will-formation of the citizens themselves." I share the concern for political opinion- and will-formation. Surely, though, Habermas's formulation collapses too much into the distinction of civic from ethnic nationalism. As we have seen, the production of similarity among members of the nation (or in their cultural practices) need not derive from either civic-political or ethnic-national sources (indeed, if it is in some sense ethnic, it is certainly not ancient or prepolitical; see Calhoun 2001).

The Public Sphere and Solidarity

What kind of solidarity does a democratic Europe require?[8] The public sphere—and thus participatory conceptions of citizenship—is only one of several ways in which solidarity can be created. Families, communities, bureaucracies, markets, movements, and ethnic nationalism are among the others.

All of these are arenas of social participation: all are institutional forms in which the members of a society may be joined together with each other. Exclusion from them is among the most basic definitions of alienation from contemporary societies. Among the various forms of social solidarity, though, the public sphere is distinctive because it is created and reproduced through discourse.[9] It is not primarily a matter of unconscious inheritance, of power relations, or of the usually invisible relationships forged as a by-product of industrial production and market exchanges. People talk in families, communities, and workplaces, of course, but the public sphere exists uniquely in, through, and for talk. It also consists specifically of talk about other social arrangements, including but not limited to actions the state might take. The stakes of theories and analyses of the public sphere, therefore, concern the extent to which communication can be influential in producing or reshaping social solidarity and producing or reshaping the social imaginary that guides participation in other forms of integration.

The classic eighteenth-century ideas of the public sphere saw it as a dimension of civil society, but one that could orient itself toward and potentially steer the state. In this sense, the public sphere did not appear as itself a self-organizing form of social solidarity, though another crucial part of civil society—the market (or economic system)—did. Rather than a form of solidarity, the public sphere was seen as a mechanism for influencing the state. Civil society provided a basis for the public sphere through nurturing individual autonomy. But the public sphere did not steer civil society directly; it influenced the state. The implication, then, was that social integration was accomplished either by power (the state) or by self-regulating systems (the economy). If citizens were to have the possibility of collective choice, they had to act on the state (which could in turn act on the economy—though too much of this might constitute a problematic dedifferentiation of spheres). Not developed in this account was the possibility that the public sphere is effective not only through informing state policy, but through forming culture; that through exercise of social imagination and forging of social relationships, the public sphere could constitute a form of social solidarity.

Publics are self-organizing fields of discourse in which participation is not based primarily on personal connections and is at least in principle open to strangers (Warner 2002). A public sphere comprises an indefinite number of more or less overlapping publics, some ephemeral, some enduring, and some shaped by struggle against the dominant organization of others. Engagement in public life establishes social solidarity partly through enhancing the significance of particular categorical identities, and partly through facilitating the creation of direct social relations. Beyond this, however, the engagement of people with each other in public is itself a form of social solidarity. This engagement includes but is not limited to rational-critical discourse about affairs of common concern. Communication in public also informs the sharing of social imaginaries, ways of understanding social life that are themselves constitutive for it. Both culture and identity are created partly in public action and interaction. An element of reasoned reflection, however, is crucial to the idea of choice as a dimension of this form of solidarity, to the distinction of public culture from simple expression of preexisting identity.

Collective subjects are formed in part in the public sphere, and it accomplishes integration at the same time that it informs choice. Of course other dimensions of identity and integration matter, but they are not discrete prior conditions. Just as it is not the case that individuals are fully formed in private life before they enter the public sphere, so peoples are not formed entirely in processes prior to political discourse. Wars, state-making, markets, and the reproduction of culture all matter and are all at least partially outside the determination of public life. But because public life also matters for the very constitution of society, there are limits to proceeding in the manner of much political and legal theory by saying, "Here is a society, how should it be governed?" As Guéhenno (1998, 1) says, "the democratic debate has focused on the distribution and use of power within a given community rather than on the definition of the community." Likewise, the democratization of Europe should not be deferred, made to wait on its integration, as though internal coherence and external boundaries must be settled first; it needs to be considered as part of the same constitutional process. And here, the question of whether Europe needs a formal, written constitution needs to be considered in relation to this broader process of creating and reshaping society itself.

Activity in the public sphere is not only about steering a state distinct from society; it is about constituting society. The intense public discourse that was part of the English and French revolutions was constitutive of collective identity at the same time as steering capacities. So too was that associated with the social question and the rise of labor movements in the nineteenth century (not only within labor ranks but also in broader public debates over inequality, rights, and social participation). The same is true today with public debates over immigration, globalization, and indeed the EU itself. They help—or at least can help—to achieve integration through democratic communication and collective action, rather than in advance of it.

Rather than treating cultures or peoples as unitary wholes, we should see such identities as shaped by a variety of contradictions, differences, overlaps, and partial disengagements. Such identities become effective not only as commonalities in relation to others but as semi-autonomous fields with their own hierarchies of exalted and denigrated membership and their own distinctive forms of cultural capital. Similarity, thus, is not the only form of cultural bond. And it is misleading to treat European or national collective identities as simply matters of commonality, thicker the more similarities they involve. Culture also frames inequalities, and individuals are embedded in culture for a sense of their specificity and differences. European integration implies the production of a plurality of new and actively created but equally European identities. That it is not limited to reconciliation of preexisting identities is important not only in relation to member nations but also to immigrants from outside Europe.

Advocates for "preexisting" identities are skeptical. Thus Anthony Smith (1995a, 131) argues that "if 'nationalism is love' . . . a passion that demands overwhelming commitment, the abstraction of 'Europe' competes on unequal terms with the tangibility and 'rootedness' of each nation." It is precisely in response to such nationalist ideas of solidarity that Habermas proposes his notion of "constitutional patriotism" in which citizens would be loyal to institutions and procedures (1997; see also 1996, 1998). This would involve only a thin notion of collective identity, and we need to ask whether that offers solidarity enough to underwrite both effective internal participation in a polity and collective response to its external engagements. Indeed, Habermas (1998, 117) himself has asked "whether there exists a functional equivalent for the fusion of the nation of citizens with the ethnic nation."

Habermas hopes the public sphere will produce a rational agreement that can take the place of preestablished culture or mere struggle over interests as the basis for political unity. He focuses especially on what would make unity legitimate and emphasizes fairness criteria embodied in a formal constitution.[10] Citizens ought rationally to choose such a constitution because it will offer procedural justice. Habermas seems to avoid an appeal to solidarity in this context because it appears too much like an appeal to preestablished culture or sentimental attachment (and he is influenced especially by discussions of German identity). But a greater emphasis on solidarity would be consistent with his larger project, and it would avoid a limit common to approaches to European integration that focus primarily on interests and identity.[11] These typically foreclose attention to capacities for choice and collective construction of social forms. In addition to positing prior interests or prior identities, they reduce the process of choice to a more or less technical implementation of that which "must be done" and reduce public debate to ratification by plebiscites.

Solidarity and democracy may both be better advanced by participation in collective struggle for a better society than by appeal to an already realized identity or an already achieved constitution. A purely formal constitutional alternative deals poorly with political belonging, assimilating solidarity too directly to legitimacy, and treats culture and social actors as essentially formed in advance of participation in the public sphere. A strong account of membership cannot be derived solely from loyalty attendant on the belief that one has chosen the best available governmental system (conversely, evaluations of what is best are apt to be heavily influenced by what seems to be one's own). Choice takes place not in the abstract but in the historically and culturally specific contexts of actual social life, among people already constituted both individually and collectively in relationships to each other. Habermas's arguments on constitutional patriotism thus seem to reflect less commitment to location of immanent potential for social change in existing historical circumstances than does some of his other work.[12] They also, I think, underestimate the solidarity-forming potential of the same public sphere that he celebrates for its contributions to reason.

Public discourse is central to any capacity for ordinary people to exercise choice over and participate in the construction of the institutions under which they live together. It does not replace elections, or social movements, or litigation, but it informs each. Or at least, it potentially informs each. The public sphere is not simply present or absent but variable in extent, liveliness, critical reason, internal compartmentalization or plurality, and efficacy.

■ Place and Media

How a European public sphere might flourish and whether it can help forge a democratic solidarity are questions about the spatial and social organization of communication. Europe has been "mapped" not just by EU membership, state borders, and market reach, but also by communicative relations. These have been changed over time in a variety of projects and institutions. Newspapers and later broadcast media, linguistic standardization, state schools, and curricular reforms all played a role in organizing communication at the level of the nation-state. Indeed, road networks, railways, and expanding markets (and market towns) also shaped and enabled communication as well as the exchange of goods. The new and largely national patterns displaced older ones such as those of the church, Latin as a lingua franca, and pilgrimage routes.

In contemporary Europe, nations continue to organize a great deal of public communication. The growth of English as an international language has not superseded national languages as the primary means of communication in everyday life and politics. European newspapers are puny beside the major national dailies. For the most part, magazine journalism also either remains national or partially transcends that into global form (like the *Economist,* which uses the slogan "Business knows no boundaries. Neither do we."). How widely languages are spoken (and read) outside the national context shapes how effectively print and broadcast media both can be exported and can potentially be organized on a transnational basis. It is worth noting, though, that even English media are minimally organized on a European scale; they are national (in the fuzzy sense that includes encompassing Britain and Scotland or England separately) or they are global. Likewise, Spanish publications sell outside of Spain, but mostly in Latin America, not other parts of Europe.

Educational systems, which are not only preparation for communication but also institutions of communication, remain basically national. Universities played a crucial role in the historical growth of both national consciousness and public spheres. It is striking that there is no European university but instead only exchanges of students among national (and to some extent subnational) institutions and the building of specialized scientific centers. The European University Institute in Florence is important but is more a cross between exchange program and research center than a university. British institutions, perhaps most prominently the London School of Economics, play a central role as places of connection, partly because of the status of English. Insead and a variety of less famous European business schools help to forge an increasingly integrated international (but largely European) business elite, but their graduates do not dominate in any national business networks, nor are companies and business communications primarily organized on a European scale.

In both print media and educational institutions, there are both increasing content about Europe and increasing similarity across nations in the way many issues are addressed. Indeed, many opinion articles are published in multiple versions, and at least at the elite of the trade, journalists maintain an awareness of what is being reported in other countries and languages. Television has its own analogous formats. Once again, though, growing cultural similarity is not the same as the development of an arena for mutual engagement and collective discussion. In this latter sense, existing print media do not support much of a European public sphere. To the extent that they do, it is a very elite affair. It is not trivial, though, that there is increased reporting in most European countries of the public discussions that take place in others. This provides for links among democracies and provides a supportive context for transnational social movements. These latter play an important role in opinion formation on a European scale.

On environment, food safety, concerns about globalization, and threats of war, there is an incipient European public sphere. At present, this is more vibrant in links among grassroots organizations than it is effectively represented in large-scale politics in most countries or on the scale of Europe. It has an ambivalent relationship to the mainstream media, which report on it episodically but generally do not provide a forum for its own communication. It is not accidental that critiques of contemporary media, such as Pierre Bourdieu's challenge (1996) to television's "cultural fast food," are an aspect of most of these movements. Their internal communications are considerably aided, however, by the Internet. And they reach publics not only through alternative newspapers and small presses and magazines, but also through the radio.

Radio occupies an important, but somewhat neglected, niche in the European mediascape. Discussion programs flourish and are significant in reaching various minority publics—whether defined ethnically, politically, or in generational terms. Radio audiences may not be any more multilingual than readers of newspapers or viewers of TV, but radio discussions seem to tolerate a more polyglot participation. To be sure, this is seldom a matter of multiple participants each speaking a different language fluently, but rather of tolerating slightly varying creolizations of a common language. Much of this radio-public sphere is an urban (not always a national) phenomenon.

The Internet is also clearly important. It is often cited as a basis for transnational civil society and political mobilization, and there is indeed evidence for its efficacy in this regard. It is clear that the "new media" (of which it is the most prominent face) will be vital to transformations of the public sphere and that in some cases these will mark extensions. Nonetheless, caution is indicated. First, the Internet and other information technologies enable markets, powerful business organizations, cultural access, and critical discourse among ordinary people to grow. But these do not necessarily grow at the same speed or wield the same influence over institutional arrangements. The new media have so far done more to enable the global organization of business than effective social movement responses or public-sphere activity.[13] Second, while there is a good deal of talk about on-line communities, most of these are in fact discussions among categories of people who share a specific common interest (parenting, say, or religion). A good many focus on sex. The senses in which they constitute communities are very limited (though, like other categories of identification among strangers, they may offer the illusion of community). Two different phenomena coexist on the Web: (1) an easy transcendence of space that is an aid to cultural circulation and sometimes political mobilization (but also global finance) and (2) a supplement to communities that remain largely local — keeping people in touch, aiding in planning, facilitating information-sharing, empowering citizens in relation to local government. These are not necessarily opposed, though the first can be part of an overall political economy that encourages people to invest themselves more in larger-scale and longer-distance relationships, organizations, and identities.

The issue is not only the way medium influences message, but also public investment in both. A key basis for democracy in European states was the growth of national communications media (though in themselves these did not guarantee democracy, of course, even where they did contribute to sociocultural integration). Newspapers were one of the most important vehicles for building national culture and identity. During the early years of television and radio, government investment in most European countries helped to make the new media function in a national manner. Willingness to subsidize national culture and national public spheres in this way has declined substantially in recent years, though it has not yet disappeared. As state media declined and private media importance grew, a veritable industry arose to assess whether the new media were becoming effective agents of European integration (e.g., Humphries 1996; Mansell and Silverstone 1997; McQuail 2001; Schlesinger 1992a, 1992b, 1997; Venturelli 1998).

One finding of this research is that while international media consumption has grown, especially in certain kinds of entertainment programming, people still prefer to hear the news in their national languages and with attention to national content. When planes crash in Asia, the European media report the number of French — or British or Swedish — passengers, not the number of Europeans.

Media offer less specifically European public communication than might be thought. They tend either to remain contained within national or linguistic audiences or to transcend Europe into some version of the global. Consider one simple example, the origin of TV fiction programs from a study of one week in 1997 (table 10.1).

In any event, it should be clear that the media are not at present well organized to support a European public sphere. They do an increasingly good job of reporting in each European country on others, but not of underpinning a com-

TABLE 10.1 Origin of TV Fiction Programs by Country of Broadcast
(in Percentages)

Country	Domestic	US	Other EU
England	51	36	0
Germany	29	60	5
France	25	58	14
Italy	17	64	2
Spain	12	62	5

SOURCE: Bondenbjerg 2001, 7.

mon process of opinion, will, or identity-formation, and especially not a critical mutual engagement. For all the talk of the Internet, broadcast media remain the most influential. They also remain organized overwhelmingly on a national basis (except for some regional and city-specific versions).

To the extent that the beginnings of a European public sphere are visible in the media, they take three forms. First, there is the official Europe of the EU and the common affairs of its members. This receives considerable mainstream media attention, though (as elections demonstrate) it excites little public enthusiasm except in moments of rejection. It is a top-down affair in which Europe is represented to Europeans from Brussels (and to some extent Strasbourg) but there is little multidirectional flow. Second, there is an elite discursive community that is much more active in public communication, is often multilingual (on the Continent, at least), reads more and more internationally, and consists largely of leaders in business and finance, parts of higher education, the media themselves, and to some extent government. This public does indeed debate European affairs, but largely in technocratic terms and with considerable attention to specific interests. Third, there are the widely ramifying networks of activists that have most recently been visible in the antiglobalization movement but in fact join those committed to many different causes, from whole foods to human—and indeed animal—rights. Although most of these movements are global in their aims and to some extent in their ultimate scope, Europe is overrepresented among their participants. The networks among activists are densest within Europe (though not remarkably dense anywhere) and flow across borders more effectively than most (certainly more than labor). In important ways, European identities may influence the activists and grow stronger through their activism (partly because of the contrast of their own orientations to an American model). Nonetheless, it is significant that the movements have so far not developed a very active discourse about shaping Europe. Their language has either been global, with aspirations for humanity or the world, or defensive (and thus national or local).

There are links among these three versions of a European public sphere, but they are impressively distinct. Moreover, they do not, even cumulatively, offer an effective way for European citizens democratically to choose their institutions and their futures.

Print media were basic to the rise of the modern public sphere, but the capacity of media to relate strangers did not obviate the importance of local settings. Face-to-face discourse in cafés and coffee houses mediated the consumption of early newspapers and anchored the eighteenth-century ideal of bourgeois public life. Pubs and taverns figured comparably in more popular practice. Along with theaters, salons, and other institutions, these spaces shaped urban publics, and the public sphere has remained importantly rooted in cities (Sennett 1977). This kind of urban life also underpins cosmopolitanism. Of course, it is not everyone's experience of cities, and in even the most cosmopolitan of them, some residents lead very circumscribed lives, by different combinations of choice and necessity. One issue in contemporary Europe is the relationship between cosmopolitan urbanites (disproportionately but not exclusively elites) and those in both cities and smaller towns whose loyalties to particular "communities" is stronger. These locals may be connected by the Internet, but they relate quite differently to issues of difference and identity, including, not least, to European integration.

Cities have been especially important as settings in which people from different contexts enter into discourse with each other. Cities are also, however, one of the dimensions of the European public sphere that is currently being transformed. Americans are wont to romanticize the public character of European cities, dwelling perhaps more on novels written in Vienna's cafés a hundred years ago than on present-day Vienna. Nonetheless, many of Europe's cities have been distinctive in their pedestrian character and their scale. Urban centers in which people of different classes, ethnic origins, and occupations rub shoulders and enter into conversation, however, house less and less of Europe's population.[14] There is still neighborhood life in European capitals, and there are still vital urban centers and intellectual districts. But Banlieux sprawl around Paris and Greater London stretches through five counties and what were a century and a half ago still dozens of separate towns and villages. Cities have lost much of their centrality to the organization of European public life, even while they have continued to grow.

One of the challenges for the future of Europe's public sphere is to find replacements for the kind of public life that flourished in face-to-face urban relations and yet spoke to the concerns of the nation as a whole. London and Paris, for example, are both remarkably multicultural; can they become bases for discourse constitutive of a democratic Europe? There are reasons for doubt. In the first place, in a long trend, public communication became more national. What were once the great newspapers of different cities increasingly became competing national newspapers (or fell by the wayside). The connection of TV to locality was never strong. Universities that were once closely tied to the character and politics of different cities are increasingly competitors to place graduates in national labor markets. Local intellectual and professional associations have generally ceased to play a major role. In short, urban public spheres lost strength.

Nonetheless, cities remain important to the future of the European public sphere. The space-transcending media will not obviate all need for mediation in face-to-face conversation. It will be crucial that there be both environments and institutions that bring together people of different identities and interests to facilitate mutual understanding and mutual engagement in both debate and the search for viable ways of living together. This matters for Europe's absorption of non-European immigrants and for transcending a variety of differences including political-economic interests, not only nationality.

Finally, one of the biggest questions about European integration is why the borders of Europe should be important (never mind just where those borders should be). It was more plausible (if still always ultimately tendentious) for nation-states to assume territorial autonomy, the containment of citizenship and public life within the national space. Nations were never as bounded as they seemed, and a unified Europe seems destined to be much less so.

Many of the issues, mobilizations, and media that link Europeans are in fact not specific to Europe. Europeans may have specific collective interests in global economic competition (though it is always worth asking which Europeans have which interests), but it is not clear that Europe is really a single unit in the international economy. It has more coherence in the world of international aid than of international military action; but in all these cases, the issue is not simply that individual nations are stronger than the EU, or diverse among themselves, but that to find a common denominator that includes all of Europe, one is usually forced to a low enough level of coherence that parts or even all of the rest of the world come into the pattern as well.

It is not enough to look inside Europe. Any understanding of European solidarity must address the different ways in which Europeans are tied to others outside Europe. These ties will obviously differentiate among Europeans — by nation, class, industry, and involvement in social movements or concerns such as human rights or the environment. There has been a great deal of attention to how Europeans remain divided on nation-state lines. But analysts exaggerate the extent to which the issue is simply inheritance of either cultural identities or specific domestic institutional regimes. Divisions will be produced and reproduced by differential incorporation into global markets, production systems, and indeed publics. Some Europeans will minimize their investment in the internal organization of Europe and maximize their commitment to firms or other organizations operating across its borders. This may be as important a form of dual identity as that of migrants. Other Europeans will mobilize global social movement ties (or international corporate power) to challenge institutional arrangements within Europe.

Nonetheless, the EU (and its member states) put a lot of effort into their collective boundary-work. This is not merely a matter of material borders, as in resisting immigration and debating enlargement of the EU. It is also a matter of giving the EU a unitary international image, representing it as an actor in a variety of global contexts, making clear its distinctive collective values. As a whole, thus, it is committed to human rights and expresses that through conditionality in aid and trade (K. Smith 2001). It is committed to capitalist economic growth and more or less liberal policies and expresses these through its central bank, its relations to global economic institutions, and (more ambiguously) its trade policies. Precisely because of the protectionism involved in the latter, and in the Common Agricultural Policy, the EU also represents itself globally as internally conflicted.

In short, the EU has an international (that is, global or supra-European) identity as a whole greater than the sum of its parts, even if not perfectly integrated, and as an actor (Manners and Whitman 1998; Whitman 1998). It gains an identity from its material actions, from its symbolic self-representations, and from the interpretations of both by others. Its identity is thus not purely fixed by treaties or legal arrangements, not entirely a reflection of material interests, and always subject to construction and reconstruction. Like the identity of any nation or state, it is in part a matter of style as well as substance, motivation, and status.

Arguably, globalization makes it harder for any geopolitical unit to maintain external closure, and certainly for "a territorially defined political unit which is not the nation-state" (Waever 1997).[15] At the same time, it is a misleading aspect of nationalist ideology to understand and represent nation-states as internally coherent and externally bounded; this is perhaps a typical nationalist project but never a fully evident reality (Mann 1998; Calhoun 1997). The EU, like nations, derives a considerable part of its identity not from its internal arrangements and boundary maintenance but from its action and representation in the world. Globalization is thus an occasion for the construction or assertion of identity, not simply a challenge to it. As we have noted, this has long been a factor in the development of the EU, and one which over time gained ascendancy over the original project of securing Europe's internal peace through prosperity and interdependence. It appears not only in the pursuit of material prosperity but also in the representation of the EU as already prosperous and indeed powerful (if more in economic and diplomatic ways than military). Since the end of the cold war, the EU has been fairly clear also that it positions itself as in some sense comparable to the United States and Japan (Manners 2001). This is both symbolically effective and to some extent misleading, because it situates the EU in comparison to states. But of course, this also reveals that some imagine the EU as a potential state.

In Waever's terms (2000), Europe weaves its international identity out of a combination of reference to a mythic narrative of its own past and its contemporary actions. The latter are by no means limited to those of the EU itself but include all that present themselves or are seen by others as specifically European. Not only nations but corporations and social movements as well may at various points appear as European, not merely as discretely themselves. One of the implications of this is that public discourse is potentially pivotal to Europe's international identity. European discourse not only addresses European identity; it forms it partly by how it represents Europe and also by how it engages the rest of the world. Yet, in Bondenbjerg's words, "we do not have an adequate national public sphere, a European public sphere or a global forum, in which this new European or global experience can be imbedded and debated in a proper democratic way" (2001, 3).

The point has long since been made that the public sphere is not seamless and integral but composed in the relationships among multiple publics and specific arenas of discourse. This is true within nation-states, on the scale of Europe, and globally. Nothing dictates that Europe will develop a strong public sphere, but if it does, this will grow in part out of the relationship of European public discourse to that of various transregional circuits of communication. How much this happens will depend not just on whether Europeans grow more similar or share interests, but on whether a mutual engagement with other Europeans becomes a central way in which they develop and come to understand identity and interests. The search for collective voice in relation to global affairs, in other words, could drive one version of European integration. Through the public sphere, it could provide a meeting point between the foreign-policy aspirations of EU officials and the global concerns of movement activists, and it could provide an occasion for choice about what being European means, for life both inside and outside the lines on the map.

■ Conclusion

I have argued that European integration is a problem in and for democracy. It is not merely an objective process to be monitored, or a response to necessity, but at least potentially also a choice. Making that choice in a democratic way is intrinsically related to having democratic outcomes from the process of integration.

I have argued also that the process of integration is not adequately understood in terms only of the interests and identities of actors. These are both important and are very closely intertwined with each other. Neither clearly comes first before all action, and thus neither can explain all action. Both are not only subject to change but can in varying degrees be formed and reformed in public life and made the object of at least partially conscious choice. At the same time, culture itself is not only a preexisting condition but also a continually renewed product made partly in public life. And part of culture is a social imaginary that shapes the way in which we construct not only our own identities but also our senses of what is real in the world and how it should be understood. The social imaginary most common in Europe today constructs corporations (whether business firms or states) as real and unitary. So, too, nations and communities—though some fight for a different way of imagining the world in which only individuals have standing. For Europe to be altogether real to its citizens (and indeed to others) requires in part a change in the social imaginary, which public discourse can encourage or resist and shape in various ways.

For all these reasons, the public sphere is a crucial setting for the production and shaping of European integration. I have argued that it can be not only a mechanism for debate and choice but also a form of and a process for forming solidarity. That it can be, however, does not mean that it will be. There are many obstacles to the development of an effective European public sphere, including the fact that this has not been a priority for the political and economic elites who have led the integration of Europe.

In thinking about the future of the European public sphere, finally, I have argued that we ought not to project false notions about internal coherence and external closure from nationalist ideology onto our expectations for European public life and integration. Plurality and heterogeneity of identities, locations, media of communication, perspectives, and arguments are all to be expected. The public sphere is most valuable as an arena in which differences are mediated, not one simply for the expression of similarities. Moreover, it is not strictly unitary but a congeries of overlapping publics on various scales, and it is no more necessarily bounded at the edges of nations—or of Europe—than markets are. Public discourse can be more or less vital, and participation in it more or less dense, on any scale, and the relations between scales can be more or less effective. Equally, the overlaps and relationships among different dense arenas of public discourse (nations, cities, media) may be strong, or they may be so weak that sharp fault lines divide the different smaller public spheres within or from the potential larger one. If public life grows on the scale of Europe, not least of all, it is important to see this as partly a matter of how Europe fits into a more global discourse, not only how Europeanness is contained in Europe.

NOTES

I am grateful for helpful comments from Kevin Moore and Mabel Berezin.

1. The public sphere is not only translocal but "metatopical," in Charles Taylor's term (1995, 263). While there are a variety of topically specific public discussions (as of place-specific public gatherings), the public sphere "knits together a plurality of such spaces into one larger space of nonassembly." Parts of the same discussion circu-

late through a variety of topical or spatial locations but derive some of their significance from being related to the larger sphere of circulation (just as the market, in a national or global sense, knits together transactions in various marketplaces but is a phenomenon of a different order). See also Calhoun 1988.

2. I have discussed this further in connection to the ideal of cosmopolitan democracy in Calhoun forthcoming. But see also the various defenses of cosmopolitanism in Archibugi forthcoming. A key motivation for relying on only a thin notion of identity is the perception that governmental (or quasi-governmental) power is organized at a variety of levels, and therefore democracy might helpfully flourish at several levels as well, not be overwhelmingly organized in terms of nation-states. See Held 1995.

3. Three options roughly equivalent to loyalty, exit, and voice; see Hirschman (1970). Of course, there are a variety of small modes of struggle that do not involve much open voice, of refusals of participation that do not amount to complete exit, and of actual participation under duress that falls short of loyalty. And it is worth recalling that Hirschman's book focuses on responses to decline in states, organizations, and firms. It does not address equally the kind of cathexes that bind people into thriving social systems, the investment in personal projects that connects individual choices and desires to the reproduction of social fields that make those projects possible (whether or not they allow them to succeed); see Bourdieu 1990.

4. The notion of ***social imaginary*** is most associated with Cornelius Castoriadis (1998), who established the extent to which social reality depends on production and reproduction in imagination. Benedict Anderson (1991) developed a partially similar account of the material and social processes underpinning the role of imagination in forging nations. For further considerations, see Gaonkar 2002.

5. Arendt's account (1958, 1990) of public space as an arena for speech that is both world-disclosing and world-making may be more helpful in this regard (and one need not adopt her denigration of the merely "social" — i.e., material and necessary — to see this). See also Taylor 1989 on the ways in which the modern notion of self includes potential for self-reformation, including wanting to have better wants.

6. See Streeck 2001 for a compelling argument on whose interests were served and how power relations and forms of citizenship were altered by the combination of an integrated European market with fragmented sovereignty and national citizenship.

7. On the rare occasions when the Eurobarometer or similar surveys gather systematic subnational data across Europe, these suggest interesting variation. EB44 was the last time the sample size was large enough to present intranational patterns of regional variation on such important questions as how much and how often respondents feel "European," or whether they think their countries' membership of the EU is a good or bad thing. Variation within some countries exceeded that among all countries, and it was considerable in most countries. All but two member states had at least one region where more than half of respondents thought their country's membership of the EU was a good thing. At the same time, all but three had at least one region where most residents thought their country's membership was a bad thing (see discussion in Manners 2001). One should not exaggerate the level of intranational dif-

ference. It did not exceed the international difference in 1996 for any country other than Finland, but national means do mask great differences—a point that merits further study. Most analyses of Eurobarometer surveys simply point to national differences. Indeed, the Eurobarometers are so insistently organized in terms of nations as units of analysis that they partially constitute a ritual reaffirmation of the "Europe of the nations" theme. They reinforce the notion that European integration is a matter of the coincidence or difference of national opinions.

8. In the context of European politics, the term *solidarity* is often narrowed to refer to certain specific social policies, especially those designed to reduce inequalities among groups, rather than being used in its more basic Durkheimian sense. Though terms like *participation* are used, the conception is technocratic and top-down; see S. Cohen 2000.

9. Elsewhere (Calhoun 2002) I have tried to lay out more formally different types of social integration, especially functional or autopoetic systems, power or domination, categorical identities such as nation, common culture (which is not the same thing as common identities), networks of directly interpersonal relationships, and public communication. The list could be longer and more fine-grained. The important point here is that all usually operate in some degree in any actual case, but with varying implications.

10. He poses this argument specifically in response to Dieter Grimm's suggestion (1997) that a constitution might be a mistake because it would raise expectations for goods Europe would not be able to deliver and thus imperil legitimacy.

11. Schmitter (2001) offers proposals to increase "sense of belonging" or commitment by increasing formal political and economic rights and participation at the EU level—e.g., referenda, separation of EU from national elections, transfer payments.

12. In general, Habermas's work since the late 1970s has relied much less heavily on the idea of immanent critique than his earlier work did. Instead, transcendental or evolutionary arguments (e.g., from the nature of communicative action) predominate.

13. Not only has the Internet been positively enabling to business, but its commercialization has also been disabling to other users. The proportion of Internet activity that is altogether amateur, organized by nonprofit organizations, social movements, or hobbyists, has been in decline for years. The costs of gaining attention for a new Web site have risen. Type almost anything into a search engine, and most of the first sites identified will be commercial.

14. This erosion is a trend at least as old as Hausman's attack on the Paris *quartiers*; see Harvey 1985. It has gained pace in the last few years, propelled forward by neoliberal policies and shifting relations among public authorities on different scales.

15. The key distinction between the EU and a nation-state, in Waever's view, would seem to be that the EU lacks sovereignty. Others point to somewhat different definitions, such as Weber's idea of maintaining a monopoly of legitimate violence. At the same time, one of the useful themes of recent analyses of "cosmopolitan democracy" (e.g., Held 1995) has been the extent to which states—even classically recognized ones—are increasingly subject to shared sovereignty.

References

Abbott, Andrew. 2001. "Temporality and Process in Social Life." In *Time Matters: On Theory and Method,* by Andrew Abbott, 209–39. Chicago: University of Chicago Press.

Abrams, Dominic, and Michael A. Hogg. 1990. "An Introduction to the Social Identity Approach." In *Social Identity Theory: Constructive and Critical Advances,* ed. Dominic Abrams and Michael A. Hogg, 1–9. New York: Springer-Verlag.

Abu-Lughod, Lila. 1991. "Writing against Culture." In *Recapturing Anthropology: Working in the Present,* ed. Richard G. Fox, 137–62. Santa Fe, NM: School of American Research Press.

Adelman, Jeremy, and Stephen Aron. 1999. "From Borderlands to Borders: Empires, Nation-States, and the Peoples in Between in North American History." *American Historical Review* 104 (3): 814–41.

Agnew, John A. 1994. "The Territorial Trap: The Geographical Assumptions of International Relations Theory." *Review of International Political Economy* 1 (1): 53–80.

———. 1995. "The Rhetoric of Regionalism: The Northern League in Italian Politics." *Transactions of the Institute of British Geographers* 20:156–72.

———. 1999. "Mapping Political Power beyond State Boundaries: Territory, Identity, and Movement in World Politics." *Millennium* 28 (3): 499–521.

———. 2001. "How Many Europes? The European Union, Eastward Enlargement, and Uneven Development." *European Urban and Regional Studies* 8 (1): 29–38.

———. 2002. *Place and Politics in Modern Italy.* Chicago: University of Chicago Press.

Agnew, John A., and Carlo Brusa. 1999. "New Rules for National Identity? The Northern League and Political Identity in Contemporary Northern Italy." *National Identities* 1:117–33.

Agnew, John A., and Stuart Corbridge. 1995. *Mastering Space: Hegemony, Territory, and International Political Economy.* New York: Routledge.

Alexander, Jeffrey A. 1997. "The Paradoxes of Civil Society." *International Sociology* 12:115–33.

———. 1998. *Real Civil Societies: Dilemmas of Institutionalization.* Thousand Oaks, CA: Sage.

Alférez, Antonio. 1986. *Cuarto poder en España: La prensa desde la ley fraga 1966.* Barcelona: Plaza e Janés.

Allen, Beverly, and Mary Russo, eds. 1997. *Revisioning Italy: National Identity and Global Culture.* Minneapolis: University of Minnesota Press.

Allen, John. 1998. "Spatial Assemblages of Power: From Domination to Empowerment." In *Human Geography Today,* ed. Doreen Massey, John Allen, and Philip Sarre, 194–218. Cambridge, MA: Polity Press.

Álvarez-Miranda, Berta. 1996. *El sur de Europa y la adhesión a la comunidad.* Madrid: Centro de Investigaciones Sociológicas.

Amit-Talai, Vered, and Helena Wulff, eds. 1995. *Youth Cultures: A Cross-Cultural Perspective.* London: Routledge.

Andersen, Svein, and Kjell Eliassen. 1996. "Introduction: Dilemmas, Contradictions, and the Future of European Democracy." In *The European Union: How Democratic Is It?* ed. S. Andersen and K. Eliassen, 1–11. Thousand Oaks, CA: Sage.

Anderson, Benedict. 1991. *Imagined Communities: Reflections on the Origin and Spread of Nationalism.* 2d ed. London: Verso.

Anderson, Elijah. 1994. *Streetwise.* Chicago: University of Chicago Press.

Anderson, Peter J., and Anthony Weymouth. 1999. *Insulting the Public? The British Press and the European Union.* London: Longman.

Ansell, Christopher K. Forthcoming. "Restructuring Authority and Territoriality: Europe and the United States Compared." In *Restructuring Territoriality: Europe and the United States Compared,* ed. Christopher K. Ansell and Giuseppe DiPalma. Cambridge: Cambridge University Press.

Appadurai, Arjun. 1996. *Modernity at Large: Cultural Dimensions of Globalisation.* Minneapolis: University of Minnesota Press.

Archibugi, Daniele. Forthcoming. *Debating Cosmopolitics.* London: Verso.

Arendt, Hannah. 1958. *The Human Condition.* Chicago: University of Chicago Press.

———. 1986. *The Origins of Totalitarianism.* London: André Deutsch.

———. 1990. *On Revolution.* London: Penguin.

Aretin, Karl Otmar von. 1993. *Das Alte Reich.* Vol. 1. Stuttgart: Klett-Cotta.

Augustine. 1972. *The City of God.* Ed. D. Knowles. Trans. H. Bettenson. Harmondsworth: Penguin.

Ausländerbeauftragte Berlin. 1992. *Kulturübergreifende Arbeit in ganz Berlin.* Berlin: Die Ausländerbeauftragte des Senats.

Averintsev, Sergei. 1991. "The Idea of Holy Russia." In *Russia and Europe,* ed. Paul Dukes, 10–23. London: Collins and Brown.

Axelrod, Robert. 1997. *The Complexity of Cooperation: Agent-Based Models of Competition and Collaboration.* Princeton, NJ: Princeton University Press.

Axford, Barry. 1995. *The Global System: Economics, Politics, and Culture.* Cambridge, MA: Polity Press.

Axtmann, Roland. 1993. "The Formation of the Modern State: The Debate in the Social Sciences." In *National Histories and European History,* ed. Mary Fulbrook, 21–45. London: UCL Press.

Axtmann, Roland, and Robert Grant. 2000. "Living in a Global World." In *Issues in International Relations,* ed. Trevor Salmon, 25–54. London: Routledge.

Babb, Sarah. 1996. "Frame Resonance in the US Labor Movement, 1866 to 1886." *American Sociological Review* 61 (6): 1033–53.

Bader, Veit. 1995. "Citizenship and Exclusion: Radical Democracy, Community, and Justice; Or, What Is Wrong with Communitarianism?" *Political Theory* 23:211–46.

Baldwin, Peter. 1999. *Contagion and the State in Europe, 1830–1930.* Cambridge: Cambridge University Press.

Baldwin-Edwards, Martin, and Martin Schain, eds. 1994. "The Politics of Immigration in Western Europe." Special edition of *West European Politics* 17 (2).

Balkır, Canan, and Alan M. Williams, eds. 1993. *Turkey and Europe*. London: Pinter Publishers.

Barber, Benjamin. 1995. *Jihad versus McWorld*. New York: Times Books.

———. 1998. *A Place for Us: How to Make Society Civil and Democracy Strong*. New York: Hill and Wang.

Barlow, Max. 1991. *Metropolitan Government*. New York: Routledge.

Barraclough, Richard. 1998. "Umberto Bossi: Charisma, Personality, and Leadership." *Modern Italy* 3:263–69.

Barzini, Luigi. 1983. *The Impossible Europeans*. London: Weidenfeld and Nicholson.

Basch, Linda, G. Nina Schiller, and S. Cristina Blanc. 1997. *Nations Unbound: Transnational Projects, Postcolonial Predicaments, and Deterritorialized Nation-States*. 4th ed. Langhorne, PA: Gordon and Breach.

Bauböck, Rainer. 1994. *Transnational Citizenship*. Aldershot, Eng.: Edward Elgar.

———. 1998. "Differentiating Citizenship." Paper presented at the conference "Citizenship and Cosmopolitanism," University of Wisconsin, Madison, November 6–8.

———. 2000. "European Integration and the Politics of Identity." Public lecture at the Austrian Permanent Representation at the European Union, Brussels, June 7.

Bauer, Otto. 1924. *Die Nationalitätenfrage und die Sozialdemokratie* (The nationality question and social democracy). Vienna.

Bauman, Zygmunt. 1990. "Modernity and Ambivalence." In *Global Culture: Nationalism, Globalization, and Modernity*, ed. Mike Featherstone, 143–69. London: Sage.

Baumann, Gerd. 1996. *Contesting Culture: Discourses of Identity in Multi-Ethnic London*. Cambridge: Cambridge University Press.

Bentham, Jeremy. 1982. *An Introduction to the Principles of Morals and Legislation*. Ed. J. H. Burns and H. L. A. Hart. London: Methuen.

Berdoulay, Vincent, and J. Nicholas Entrikin. 1998. "Lieu et sujet: Perspectives théoretiques." *L'espace géographique* 27:111–21.

Berezin, Mabel. 1997. *Making the Fascist Self: The Political Culture of Inter-War Italy*. Ithaca, NY: Cornell University Press.

———. 1999. "Emotions Unbound: Feeling Political Incorporation in the New Europe." Paper presented to the American Sociological Association Meetings, Chicago.

———. 2000. "The Euro Is More Than Money: Converting Currency, Exchanging Identity, and Selling Citizenship in Post-Maastricht Europe." *Policy Newsletter*, Center for Economy and Society, University of Michigan Business School, vol. 1, issue 1. www.bus.umich.edu/cse/.

———. 2002. "Secure States: Towards a Political Sociology of Emotion." In *Sociological Review Monograph*, ed. Jack Barbalet, 33–52. London: Blackwell.

Berger, V. 1994. *Jurisprudence de la Cour européenne des droits de l'homme*. 4th ed. Paris: Sirey.

Beriss, David. 1990. "Scarves, Schools, and Segregation: The Foulard Affair." *French Politics and Society* 8:1–15.

Berlin: Open City, the Guide. 1999. Berlin: Nicolaische Verlagsbuchhandlung.

Berliner Taschenwegweiserin für Mädchen. 1995. Berlin: Verband für sozial-kulturelle Arbeit.

Bezirksamt Kreuzberg. 1990. *Jugendarbeit in Kreuzberg: Dokumentation der Entwicklung.* Berlin: Bezirksamt Kreuzberg von Berlin.

Bhabha, Homi K. 1994a. "Between Identities." In *Migration and Identity,* ed. Rina Benmayor and Andor Skotnes. Oxford: Oxford University Press.

———. 1994b. *The Location of Culture.* London: Routledge.

Bhabha, Jacqueline. 1999. "Belonging in Europe: Citizenship and Post-National Rights." *International Social Science Journal* 159 (March): 11–23.

Biersteker, Thomas, and Cynthia Weber, eds. 1996. *State Sovereignty as Social Construct.* Cambridge: Cambridge University Press.

Biggs, Michael. 1999. "Putting the State on the Map: Cartography, Territory, and European State Formation." *Comparative Studies in Society and History* 41 (2): 374–405.

Biorcio, Roberto. 1997. *La Padania promessa.* Milan: Il Saggiatore.

Bjørgo, Tore. 1997. "'The Invaders,' 'the Traitors,' and 'the Resistance Movement': The Extreme Right's Conceptualisation of Opponents and Self in Scandinavia." In Modood and Werbner 1997, 54–72.

Bondenbjerg, Ib. 2001. "European Media, Cultural Integration, and Globalization." *Nordicom Review* 22 (1): 7. www.lboro.ac.uk/research.changing.media/Nordicom-2.htm.

Borneman, John. 1997. "State, Territory, and National Identity Formation in the Two Berlins, 1945–1995." In Gupta and Ferguson 1997, 93–117.

Borneman, John, and Nick Fowler. 1997. "Europeanization." *Annual Review of Anthropology* 26:487–514.

Borri, Michele. 1994. "National Identities and Attitudes towards Europe: A Comparative Study." Paper presented to the British Council–University of Silesia Conference "Culture and Identity," Cieszyn, Poland, September 22–25.

Bourdieu, Pierre. 1977. *Outline of a Theory of Practice.* Cambridge: Cambridge University Press.

———. 1985. "The Social Space and the Genesis of Groups." *Theory and Society* 14 (6): 723–44.

———. 1989. "Social Space and Symbolic Power." *Sociological Theory* 7 (1) (spring): 14–25.

———. 1990. *The Logic of Practice.* Stanford, CA: Stanford University Press.

———. 1996. *On Television.* New York: New Press, 1998.

Braudel, Ferdinand. 1980. *On History.* Trans. S. Matthews. Chicago: University of Chicago Press.

Brenner, Neil. 2001. "Entrepreneurial Cities, 'Glocalizing' States, and the New Politics of Scale: Rethinking the Political Geographies of Urban Governance in Western Europe." Working Paper 76a/76b, Center for European Studies, Harvard University.

Brenner, Neil, Bob Jessop, Martin Jones, and Gordon MacLeod, eds. 2003. *State/Space: A Reader.* Oxford: Blackwell.

Brewin, Christopher. 1997. "Society as a Kind of Community: Communitarian Voting with Equal Rights for Individuals in the European Union." In Modood and Werbner 1997, 223–39.

Bröder, Friedrich J. 1976. ***Presse und Politik: Demokratie und Gesellschaft im Spiegel politischer Kommentare der*** Frankfürter Allgemeinen Zeitung, ***der*** Welt ***und der*** Süddeutschen Zeitung. Erlangen: Palm und Enke.

Bröskamp, Bernd. 1994. ***Körperliche Fremdheit: Zum Problem der interkulturellen Begegnung im Sport.*** Sankt Augustin: Academia Verlag.

Brubaker, Rogers. 1992. ***Citizenship and Nationhood in France and Germany.*** Cambridge: Harvard University Press.

Brubaker, Rogers, and Fred Cooper. 2000. "Beyond Identity." ***Theory and Society*** 29:1–47.

Brubaker, Rogers, and David D. Laitin. 1998. "Ethnic and Nationalist Violence." ***Annual Review of Sociology*** 24:423–52.

Brunet, Roger. 1989. ***Les villes "européennes."*** Paris: DATAR.

Bruneteau, Bernard. 2000. "The Construction of Europe and the Concept of the Nation State." ***Contemporary European History*** 9 (2): 245–60.

Bruni, Frank. 2002. "Persistent Drop in Fertility Reshapes Europe's Future." ***New York Times,*** 26 December, 1, A10.

Buchanan, James, and Gordon Tullock. 1962. ***The Calculus of Consent.*** Chapel Hill: University of North Carolina Press.

Bull, Hedley. 1977. ***The Anarchical Society.*** New York: Columbia University Press.

Bullmann, Udo. 1991. ***Kommunale Strategien gegen Massenarbeitslosigkeit: Ein Einstieg in die sozialökologische Erneuerung.*** Opladen: Leske und Budrich.

Çağlar, Ayşe S. 1997. "Hyphenated Identities and the Limits of 'Culture.'" In Modood and Werbner 1997, 169–85.

———. 1998. "Popular Culture, Marginality, and Institutional Incorporation: German-Turkish Rap and Turkish Pop in Berlin." ***Cultural Dynamics*** 10 (3): 243–61.

Calhoun, Craig. 1988. "Populist Politics, Communications Media, and Large Scale Social Integration," ***Sociological Theory*** 6 (2): 219–41.

———. 1992. "The Infrastructure of Modernity: Indirect Relationships, Information Technology, and Social Integration." In ***Social Change and Modernity,*** ed. Hans Haferkamp and Neil J. Smelser, 205–36. Berkeley: University of California Press.

———. 1995. ***Critical Social Theory.*** Cambridge, MA: Blackwell.

———. 1997. ***Nationalism.*** Minneapolis: University of Minnesota Press.

———. 1998a. "Community without Propinquity Revisited: Communication Technology and the Transformation of the Urban Public Sphere." ***Sociological Inquiry*** 68 (3): 373–97.

———. 1998b. "Whose Public, Which Good?" In ***Private Action and the Public Good,*** ed. Walter Powell and Elizabeth Clemens, 20–35. New Haven, CT: Yale University Press.

———. 1999. "Nationalism, Political Community, and the Representation of Society; Or, Why Feeling at Home Is Not a Substitute for Public Space." ***European Journal of Social Theory*** 2 (2): 217–31.

———. 2001. "Identity and Plurality in the Conceptualization of Europe." In ***Constructing Europe's Identity: The External Dimension,*** ed. Lars-Eric Cederman, 35–56. Boulder, CO: Lynne Reiner.

———. 2002. "Imagining Solidarity: Cosmopolitanism, Constitutional Patriotism, and the Public Sphere." ***Public Culture*** 14 (1): 147–72.

———. Forthcoming. "The Class Consciousness of Frequent Travelers: Toward a Critique of Actually Existing Cosmopolitanism." In Archibugi forthcoming.
———, ed. 1992. *Habermas and the Public Sphere.* Cambridge: MIT Press.
Cartocci, Roberto. 1994. *Fra Lega e Chiesa: L'Italia in cerca di integrazione.* Bologna: Il Mulino.
Casey, Edward S. 1997. *The Fate of Place: A Philosophical History.* Berkeley: University of California Press.
Castells, Manuel. 1977. *The Urban Question.* Cambridge: MIT Press.
———. 1997. *The Power of Identity.* Oxford: Blackwell.
———. 1998. *End of Millennium.* Malden, MA: Blackwell.
Castoriadis, Cornelius. 1998. *The Imaginary Institution of Society.* Cambridge: MIT Press.
Certeau, Michel de. 1984. *The Practice of Everyday Life.* Berkeley: University of California Press.
Charter of Fundamental Rights of the European Union. 2000. *Official Journal of the European Communities,* 2000/C 364/01 (18 December).
Cheshire, Paul, and Ian Gordon. 1996. "Territorial Competition and the Predictability of Collective Inaction." *International Journal of Urban and Regional Research* 20 (3): 383–99.
Christiansen, Thomas, Knud Erik Jorgensen, and Antje Wiener, eds. 1999. "The Social Construction of Europe." *Journal of European Public Policy* (special issue) 6 (4): 527.
Clout, Hugo, ed. 1981. *Regional Development in Western Europe.* 2d ed. New York: Wiley.
Cohen, Jean L. 1985. "Strategy or Identity: New Theoretical Paradigms and Contemporary Social Movements." *Social Research* 52:663–716.
Cohen, Robin. 1997. *Global Diasporas: An Introduction.* Seattle: University of Washington Press.
Cohen, Roger. 2000. "Six Nations Seem Ready to Ease Up on Austria." *New York Times,* 10 May.
Cohen, Stanley. 1988. *Against Criminology.* Oxford: Transaction Books.
Cohen, Sue. 2000. "Social Solidarity in the Delors Period: Barriers to Participation." In *Democratizing the European Union,* ed. C. Hoskyns and M. Newman, 12–38. Manchester: Manchester University Press.
Cole, John, and Francis Cole. 1993. *The Geography of the European Community.* London: Routledge.
Coleman, James. 1989. *Foundations of Social Theory.* Cambridge: Harvard University Press.
Colley, Linda. 1992. *Britons: Forging the Nation, 1707–1837.* New Haven, CT: Yale University Press.
Confino, Alon. 1997. *The Nation as a Local Metaphor: Württemberg, Imperial Germany, and National Memory, 1870–1918.* Chapel Hill: University of North Carolina Press.
Connolly, William E. 1991. *Identity/Difference.* Ithaca, NY: Cornell University Press.
Cooke, Philip, and Kevin Morgan. 1995. *The Associational Economy.* New York: Oxford University Press.
Corse, Sarah. 1997. *Nationalism and Literature.* Cambridge: Cambridge University Press.
Creveld, Martin van. 1999. *The Rise and Decline of the State.* Cambridge: Cambridge University Press.

Curry, Michael. 1996. *The Work in the World: Geographical Practice and the Written Word.* Minneapolis: University of Minnesota Press.

Daley, Suzanne. 2001. "Despite European Unity Efforts, to Most Workers There's No Country like Home." *New York Times,* 12 May, A6.

Dalton, Russell J., and Robert Duval. 1981. "The Political Environment and Foreign Policy Opinions: British Attitudes towards European Integration, 1972–1979." *British Journal of Political Science* 16:113–34.

Darnton, Robert. 2002. "A Euro State of Mind." *New York Review of Books,* 28 February, 30–32.

DATAR. 1989. *Les villes "européennes."* Paris: La Documentation Française.

Davies, Norman. 1996. *Europe: A History.* Oxford: Oxford University Press.

Davis, Roy W. 2002. "Citizenship of the Union . . . Rights for All?" *European Law Review* 27:121–37.

Dawson, Christopher. 1960. *Understanding Europe.* New York: Image Books.

Deas, Ian, and Kevin Ward. 2000. "From the 'New Localism' to the 'New Regionalism'? The Implications of Regional Development Agencies for City-Regional Relations." *Political Geography* 19:273–92.

Deflem, Mathieu, and Fred C. Pample. 1996. "The Myth of Postnational Identity: Popular Support for European Unification." *Social Forces* 75 (1): 119–43.

Delanty, Gerard. 1995. *Inventing Europe: Idea, Identity, Reality.* London: Macmillan.

Deutsch, Karl W. 1953. *Nationalism and Social Communication.* Cambridge: Technology Press, MIT.

———. 1968. *Analysis of International Relations.* Englewood Cliffs, NJ: Prentice-Hall.

Dewey, John. [1916] 1944. *Democracy and Education: An Introduction to the Philosophy of Education.* New York: Free Press.

———. 1927. *The Public and Its Problems.* Denver: A. Swallow.

De Witte, Bruno. 1993. "The European Community and Its Minorities." In *Peoples and Minorities in International Law,* ed. Catherine Brölmann, René Lefeber, and Marjoleine Zieck, 167–85. Kluwer, Netherlands: Academic Publishers.

Diamanti, Ilvo. 1997. "Nuove generazioni: L'Europa è lontana, l'Italia un po' meno." *Il Mulino* 56:46–54.

Diamond, Jared. 1997. *Guns, Germs, and Steel.* New York: Norton.

Díez-Medrano, Juan. 2003. *Framing Europe: Attitudes toward European Integration in Germany, Spain, and the United Kingdom.* Princeton, NJ: Princeton University Press.

Diop, A. M. 1994. "Structuration d'un réseau: La Jamaat-Tabligh (société pour le propagation de la foi)." *Revue européenne des migrations internationales* 10 (1): 145–57.

———. 1997. "Negotiating Religious Difference: The Opinions and Attitudes of Islamic Association in France." In Modood and Werbner 1997, 111–25.

Dotzauer, Winfried. 1998. *Die Deutschen Reichskreise, 1383–1806.* Stuttgart: Franz Steiner.

"Douze pays sortent leurs mouchoirs." 2001. *Liberation* (Paris), 31 December, 4.

Duncan, Simon, and Mark Goodwin. 1989. *The Local State and Uneven Development.* London: Polity Press.

Dunford, Mick, and Diane Perrons. 1994. "Regional Inequality, Regimes of Accumulation, and Economic Development in Contemporary Europe." *Transactions of the Institute of British Geographers* 19:163–82.

Durkheim, Emile. [1893] 1984. *The Division of Labor in Society.* New York: Free Press.

Eatwell, Roger, ed. 2000. "Far-Right in Europe: In or Out of the Cold." *Parliamentary Affairs* (special issue) 53 (3): 407–531.

Eder, Klaus, and Bernhard Giesen, eds. 2001. *European Citizenship between National Legacies and Postnational Projects.* Oxford: Oxford University Press.

Eichenberg, Richard, and Russell J. Dalton. 1993. "Europeans and the European Community: The Dynamics of Public Support for European Integration." *International Organization* 47:507–34.

Eisenschitz, Aram, and Jamie Gough. 1993. *The Politics of Local Economic Development.* New York: Macmillan.

Eisenstadt, Shmuel Noah, and Bernhard Giesen. 1995. "The Constitution of Collective Identity." *Archives européennes de sociologie* 36:72–102.

Eliot, T. S. 1962. "The Unity of European Culture." In *Notes towards the Definition of Culture,* appendix, 110–24. London: Faber and Faber.

Elliott, J. H. 1992. "A Europe of Composite Monarchies." *Past and Present* 137:48–71.

Entrikin, J. Nicholas. 1991. *The Betweenness of Place: Towards a Geography of Modernity.* Baltimore: Johns Hopkins University Press.

———. 1997. "Lieu, culture et démocratie." *Cahiers de géographie du Québec* 41:349–56.

Erlanger, Steven. 2002. "Euro Edges Past the Dollar in Victory for the Europeans." *New York Times,* 16 July.

Esping-Andersen, Gosta. 1997. "Welfare States at the End of the Century: The Impact of Labour Market, Family, and Demographic Change." In *Family, Market, and Community: Equity and Efficiency in Social Policy,* 63–80. Social Policy Studies, no 21. Paris: OECD.

European Commission. 1994. *The Citizens and the Single Market.* Brussels: European Commission.

———. 1998. "Commission Communication on the Information Strategy for the Euro." *Euro Papers,* no. 16 (February).

———. 1999. *Affirming Fundamental Rights in the European Union: A Time to Act.* Brussels: European Commission.

Evans, Peter. 1997. "The Eclipse of the State? Reflections on Stateness in an Era of Globalization." *World Politics* 50 (1): 62–87.

Fabian, Johannes. 1983. *Time and the Other: How Anthropology Makes It Object.* New York: Columbia University Press.

Fabre, Daniel. 1996. "L'ethnologue et les nations." In *L'Europe entre cultures et nations,* ed. Daniel Fabre, 99–120. Paris: Editions de la Maison des Sciences de l'Homme.

Faist, Thomas. 1998. "International Migration and Transnational Social Spaces." *Archives Européennes de Sociologie* 39 (2): 213–47.

Favell, Adrian. 1997. "Citizenship and Immigration: Pathologies of a Progressive Philosophy." *New Community* 23 (2): 173–95.

———. 2001a. "L'Européanisation ou l'émergence d'un nouveau 'champ politique': Le cas de la politique d'immigration." *Cultures et conflits* 38–39:153–87.

———. 2001b. *Philosophies of Integration: Immigration and the Idea of Citizenship in France and Britain.* 2d ed. Houndmills, Basingstoke: Palgrave.

Feldblum, Miriam. 1999. *Reconstructing Citizenship: The Politics of Nationality Reform and Immigration in Contemporary France.* Albany: State University of New York Press.

Fentress, James, and Chris Wickham. 1992. *Social Memory.* Oxford: Blackwell.

Ferry, Jean-Marc. 1991. "Pertinence du postnational." *Esprit,* November, 80–94.

———. 2001. *L'état européen.* Paris: Gallimard.

Finer, Samuel E. 1975. "State and Nation-Building in Europe: The Role of the Military." In *The Formation of National States in Western Europe,* ed. Charles Tilly, 84–163. Princeton, NJ: Princeton University Press.

Fligstein, Neil, and Iona Mara-Drita. 1996. "How to Make a Market: Reflections on the Attempt to Create a Single Market in the European Union." *American Journal of Sociology* 102 (1): 1–33.

Føllesdal, Andreas. 1998. "Survey Article: Subsidiarity." *Journal of Political Philosophy* 6:190–218.

Foner, Nancy, Rube G. Rumbaut, and Steven J. Gold. 2001. "Immigration and Immigration Research in the United States." *Items and Issues* 2 (1–2) (summer): 1–6.

Forest, Benjamin. 1995. "Taming Race: The Role of Space in Voting Rights Litigation." *Urban Geography* 16:98–111.

Fox Przeworski, Joanne. 1986. "Changing Intergovernmental Relations and Urban Economic Development." *Environment and Planning C: Government and Policy* 4 (4): 423–39.

Friedland, Roger, and Robert R. Alford. 1991. "Bringing Society Back In: Symbols, Practices, and Institutional Contradictions." In *The New Institutionalism in Organizational Analysis,* ed. Walter W. Powell and Paul J. DiMaggio, 232–63. Chicago: University of Chicago Press.

Furet, François. 1995. "Europe after Utopianism." *Journal of Democracy* 6:79–89.

Gabel, Matthew. 1998. *Interests and Integration: Market Liberalization, Public Opinion, and European Union.* Ann Arbor: University of Michigan Press.

Gal, Susan, and Gail Kligman. 2000. *The Politics of Gender after Socialism: A Comparative Historical Essay.* Princeton, NJ: Princeton University Press.

Galtung, Johan. 1969. "On the Future of the International System." In *Mankind 2000,* ed. Robert Jungk and Johan Galtung, 12–41. Oslo: Universitetsforlaget.

Gambetta, Diego. 1993. *The Sicilian Mafia: The Business of Private Protection.* Cambridge: Harvard University Press.

Gaonkar, Dilip, ed. 2002. *New Imaginaries.* Durham, NC: Duke University Press.

Garcia, Soledad, ed. 1993. *European Identity and the Search for Legitimacy.* London: Pinter.

Geertz, Clifford. 1973. *The Interpretation of Cultures.* New York: Basic Books.

Gellner, Ernest. 1983. *Nations and Nationalism.* Oxford: Blackwell.

Gierke, Otto von. 1922. *Political Theories of the Middle Age.* Trans. F. W. Maitland. Cambridge: Cambridge University Press.

Gieryn, Thomas F. 2000. "A Space for Place in Sociology." *Annual Review of Sociology* 26:463–96.

Gilroy, Paul. 1987. *There Ain't No Black in the Union Jack.* London: Hutchinson.

———. 1993. *The Black Atlantic: Modernity and Double Consciousness.* Cambridge: Harvard University Press.

Giner, Juan A. 1983. "Journalists, Mass Media, and Public Opinion in Spain, 1938–1982." In *The Press and the Rebirth of Iberian Democracy,* ed. Kenneth Maxwell, 33–55. Westport, CT: Greenwood Press.

Glacken, Clarence J. 1967. *Traces on the Rhodian Shore.* Berkeley: University of California Press.

Goddard, Victoria A., Josep R. Llobera, and Cris Shore, eds. 1994. *The Anthropology of Europe: Identity and Boundaries in Conflict.* Oxford: Berg.

Goldstein, Leslie Friedman. 2001. *Constituting Federal Sovereignty: The European Union in Comparative Context.* Baltimore: Johns Hopkins University Press.

Goodwin, Mark, and Joe Painter. 1996. "Local Governance, the Crises of Fordism, and the Changing Geographies of Regulation." *Transactions of the Institute of British Geographers* 21:635–48.

Goodwin-White, Jamie. 1998. "Where the Maps Are Not Yet Finished: A Continuing American Journey." In *The Immigration Reader: America in Multidisciplinary Perspective,* ed. David Jacobson, 415–29. Oxford: Blackwell.

Gottmann, Jean. 1975. "The Evolution of the Concept of Territory." *Social Science Information* 14 (3–4): 29–47.

Gould, Roger V. 2000. "Revenge as Sanction and Solidarity Display: An Analysis of Vendettas in Nineteenth-Century Corsica." *American Sociological Review* 65:682–704.

Graubard, Stephen, ed. 1995. "What Future for the State?" *Daedalus* 124 (2).

Griffin, Christine. 1993. *Representations of Youth: The Study of Youth and Adolescence in Britain and America.* Cambridge, Eng.: Polity Press.

Grimm, Dieter. 1997. "Does Europe Need a Constitution?" In *The Question of Europe,* ed. Peter Gowan and Perry Anderson, 239–58. London: Verso.

Guéhenno, Jean-Marie. 1998. "European Integration: The End of Politics or the Rebirth of Democracy." *Hessische Gesellschaft für Demokratie und Ökologie.* www.hgdoe.de/pol/guehenn.htm.

Guibernau, Montserrat. 1996. *Nationalisms: The Nation-State and Nationalism in the Twentieth Century.* Cambridge, Eng.: Polity Press.

Guild, Elspeth. 1996. "The Legal Framework of Citizenship of the European Union." In *Citizenship, Nationality, and Migration in Europe,* ed. David Cesarani and Mary Fulbrook, 30–54. London: Routledge.

Guizot, François. [1828] 1997. *The History of Civilization in Europe.* Trans. William Hazlitt. Ed. Larry Siedentop. London: Penguin Books.

Gupta, Akhil, and James Ferguson. 1992. "Beyond 'Culture': Space, Identity, and the Politics of Difference." *Cultural Anthropology* 7:6–23.

———, eds. 1997. *Culture, Power, Place.* Durham, NC: Duke University Press.

Guttman, Amy. 1993. "The Challenge of Multiculturalism in Political Ethics." *Philosophy and Public Affairs* 22 (3): 171–206.

Habermas, Jürgen. 1989. *Structural Transformation of the Public Sphere.* Trans. Thomas Burger. Cambridge: MIT Press.

———. 1992a. "Citizenship and National Identity: Some Reflections on the Future of Europe." *Praxis International* 12 (1): 1–19.

———. 1992b. "Citoyenneté et identité nationale." In Lenoble and Dewandre 1992, 17–39.

———. 1992c. "Further Reflections on the Public Sphere." In Calhoun 1992, *Habermas,* 421–61.

———. 1996. *Between Facts and Norms: Contributions to a Discourse Theory of Law and Democracy.* Trans. W. Rehg. Cambridge: MIT Press.

———. 1997. "Reply to Grimm." In *The Question of Europe,* ed. Peter Gowan and Perry Anderson, 259–64. London: Verso.

———. 1998. *The Inclusion of the Other.* Ed. C. Cronin and P. De Greiff. Cambridge: MIT Press.

Hahne, Ulf. 1985. *Regionalentwicklung durch Aktivierung intraregionaler Potentiale.* Schriften des Instituts für Regionalforschung der Universität Kiel. Vol. 8. Munich: Florenz.

Halbwachs, Maurice. [1925] 1975. *Les cadres sociaux de la mémoire.* Paris: Mouton.

Hall, Stuart. 1992. "The Question of Cultural Identity." In *Modernity and Its Future,* ed. Stuart Hall, David Held, and Tony McGrew, 273–316. Cambridge, Eng.: Polity Press.

Hall, Stuart, and Tony Jefferson. 1983 [1975]. *Resistance through Rituals: Youth Sub-Cultures in Post-War Britain.* London: Hutchinson.

Hall, Tim, and Phil Hubbard, eds. 1998. *The Entrepreneurial City: Geographies of Politics, Regime, and Representation.* London: Wiley.

Hannertz, Ulf. 1996. *Transnational Connections: Culture, People, Places.* London, Routledge.

Hansen, Randall, and Patrick Weil, eds. 2001. *Towards a European Nationality: Citizenship, Immigration, and Nationality Law in the European Union.* New York: St. Martin's.

Harding, Alan. 1994. "Urban Regimes and Growth Machines: Towards a Cross-National Research Agenda," *Urban Affairs Quarterly* 29 (3): 356–82.

———. 1997. "Urban Regimes in a Europe of the Cities?" *European Urban and Regional Studies* 4 (4): 291–314.

Harris, David, and John Darcy. 2001. *The European Social Charter.* 2d ed. Vol. 25 in the Procedural Aspects of International Law Monograph Series. Ardsley, NY: Transnational Publishers.

Harvey, David. 1982. *The Limits to Capital.* Chicago: University of Chicago Press.

———. 1985. *Consciousness and the Urban Experience.* Baltimore: Johns Hopkins University Press.

———. 1989a. *The Condition of Postmodernity.* Oxford: Blackwell.

———. 1989b. "From Managerialism to Entrepreneurialism: The Transformation in Urban Governance in Late Capitalism." *Geografiska Annaler* B 71 (1): 3–18.

———. 1989c. *The Urban Experience.* Baltimore: Johns Hopkins University Press.

Hebdige, Dick. 1979. *Subculture: The Meaning of Style.* London: Routledge.

Hedetoft, Ulf. 1997. "The Cultural Semiotics of 'European Identity': Between National Sentiment and the Transnational Imperative." In *Rethinking the European Union,* ed. A. Landau and R. G. Whitman, 147–70. London: Macmillan.

———. 1999. "The Nation-State Meets the World: National Identities in the Context of Transnationality and Cultural Globalization." *European Journal of Social Theory* 2 (1): 71–94.

Heinz, Werner, ed. 2000. *Stadt und Region–Kooperation oder Koordination? Ein internationaler Vergleich*. Schriften des Deutschen Instituts für Urbanistik, vol. 93. Stuttgart: Verlag W. Kohlhammer.

Heitmeyer, Wilhelm, Joachim Müller, and Helmut Schröder. 1997. *Verlockender Fundamentalismus: Türkische Jugendliche in Deutschland*. Frankfurt am Main: Suhrkamp.

Held, David. 1995. *Democracy and the Global Order: From the Modern State to Cosmopolitan Governance*. Cambridge, Eng.: Polity Press.

Held, David, Anthony McGrew, David Goldblatt, and Jonathan Perraton. 1999. *Global Transformations*. Stanford, CA: Stanford University Press.

Herzfeld, Michael. 1987. *Anthropology through the Looking-Glass: Critical Ethnography in the Margins of Europe*. Cambridge: Cambridge University Press.

Hewstone, Miles. 1986. *Understanding Attitudes to the European Community*. Cambridge: Cambridge University Press.

Hirschman, Albert. 1970. *Exit, Voice, and Loyalty*. Princeton, NJ: Princeton University Press.

Hirst, Paul. 1997. *From Statism to Pluralism*. London: UCL Press.

Hirst, Paul, and Grahame Thompson. 1999. *Globalization in Question*. 2d ed. Oxford: Polity Press.

Hobbes, Thomas. [1651] 1991. *Leviathan*. Ed. Richard Tuck. Cambridge: Cambridge University Press.

Hobsbawm, Eric. 1983. "Mass-Producing Traditions: Europe, 1870–1914." In Hobsbawm and Ranger 1983, 263–307.

Hobsbawm, Eric, and Terrence Ranger, eds. 1983. *The Invention of Tradition*. Cambridge: Cambridge University Press.

Hoeber, S. Rudolf, and James Piscatory, eds. 1997. *Transnational Religion*. Oxford: Westview.

Hoffman, Stanley. 1993. "Thoughts on the French Nation Today." *Daedalus* 122 (3): 63–79.

Hogg, Michael A., and Craig McGarty. 1990. "Self-Categorization and Social Identity." In *Social Identity Theory: Constructive and Critical Advances*, ed. D. Abrams and M. A. Hogg, 10–27. New York: Springer-Verlag.

Horrocks, David, and Eva Kolinsky, eds. 1996. *Turkish Culture in German Society Today*. Providence, RI: Berghahn Books.

Horst, Alfred Heinrich. 1996. "Generations bedingte zeithistorische Erinnerung in Deutschland: Ergebnisdokumentation einer computergestützten Imhaltsanalyse mit INTEXT: Arbeitsberichte aus dem DFG-Projekt 'Nationale Identität der Deutschen.'" *Messung und Erklärung der Veränderungsprozesse in Ost und West* 10:1–62.

Howitt, Richard. 1993. "'A World in a Grain of Sand': Towards a Reconceptualization of Geographical Scale." *Australian Geographer* 24:33–44.

———. 1998. "Scale as Relation: Musical Metaphors of Geographical Scale." *Area* 30:49–58.

Humphries, Peter J. 1996 *Mass Media and Media Policy in Western Europe*. Manchester: Manchester University Press.

Hutchings, Kimberly. 1996. *Kant, Critique, and Politics*. London: Routledge.

Ignatieff, Michael. 1994. *Blood and Belonging: Journeys into the New Nationalism*. London: Vintage.

Imig, Doug, and Sidney Tarrow. 2001. "Mapping the Europeanization of Contention: Evidence from a Quantitative Data Analysis." In *Contentious Europeans: Protest and Politics in an Emerging Polity*, ed. Doug Imig and Sidney Tarrow, 27–49. Boulder, CO: Rowman and Littlefield.

"The Immigrant Lockup." 1998. *New York Times*, 31 December, A18.

Inayatullah, Naeem, and Mark Rupert. 1994. "Hobbes, Smith, and the Problem of Mixed Ontologies." In *The Global Economy as Political Space*, ed. Stephen Rosow, Naeem Inayatullah, and Mark Rupert. Boulder, CO: Lynne Rienner.

Inglehart, Ronald. 1977. "Long Term Trends in Mass Support for European Unification." *Government and Opposition* 12:150–77.

"IPO = Instrumentum Pacis Osnabrugense." 1969. In *Instrumenta Pacis Westphalicae*, ed. Konrad Müller. 2d ed. Bern: Lang.

Jackson, Robert. 1999. "Sovereignty in World Politics: A Glance at the Conceptual and Historical Landscape." *Political Studies* 47:431–56.

Jacobson, David. 1996. *Rights across Borders: Immigration and the Decline of Citizenship*. Baltimore: Johns Hopkins University Press.

Jacobson, Norman. 1998. "The Strange Case of the Hobbesian Man." *Representations* 63:1–12.

James, William. 1956. "The Sentiment of Rationality." In *The Will to Believe and Other Essays in Popular Philosophy*, 63–110. New York: Dover.

Jessop, Bob. 1989. "Conservative Regimes and the Transition to Post-Fordism: The Cases of Great Britain and West Germany." In *Capitalist Development and Crisis Theory*, ed. M. Gottdiener and N. Komninos, 261–99. New York: St. Martin's Press.

———. 1998. "The Narrative of Enterprise and the Enterprise of Narrative: Place-Marketing and the Entrepreneurial City." In Hall and Hubbard 1998, 77–102.

———. 2000. "The Crisis of the National Spatio-Temporal Fix and the Ecological Dominance of Globalizing Capitalism." *International Journal of Urban and Regional Research* 24 (2): 323–60.

Jones, Peter. 1999. "Human Rights, Group Rights, and People's Rights." *Human Rights Quarterly* 21 (1): 80–107.

Jönsson, Christer, Sven Tägil, and Gunnar Törnqvist. 2000. *Organizing European Space*. London: Sage.

Joppke, Christian. 1999. *Immigration and the Nation-State: The United States, Germany, and Great Britain*. Oxford: Oxford University Press.

———, ed. 1998. *Challenges to the Nation State: Immigration in Western Europe and the United States*. Oxford: Oxford University Press.

Joppke, Christian, and Steven Lukes. 1999. *Multicultural Questions*. New York: Oxford University Press.

Judt, Tony. 2000. "Tale from the Vienna Woods." *New York Review of Books*, 23 March, 8–9.

Jusdanis, Gregory. 2001. *The Necessary Nation*. Princeton, NJ: Princeton University Press.

Kann, Robert A. 1974. *A History of the Habsburg Empire, 1526–1918*. Berkeley: University of California Press.

Kastoryano, Riva. 1994. "Mobilisations ethniques: Du national au transnational." *Revue européenne des migrations internationales* (special issue) 10 (1): 169–83.

———. 1996. *La France, l'Allemagne et leurs immigrés: Négocier l'identité*. Paris: Armand Colin.

———. 1997. "Participation transnationale et citoyenneté: Les immigrés dans l'Union européenne." *Cultures et conflits* 28:59–75.

———. 2002. *Negotiating Identities: States and Immigrants in France and Germany*. Princeton, NJ: Princeton University Press.

———, ed. 1998. *Quelle identité pour l'Europe? Le multiculturalisme à l'épreuve*. Paris: Presses de Sciences-Po.

Kaya, Ayhan. 2001. *"Sicher in Kreuzberg": Constructing Diasporas: Turkish Hip-Hop Youth in Berlin*. Berlin: Transcript Verlag.

Keane, John. 1995. "Nations, Nationalism, and European Citizens." In *Notions of Nationalism*, ed. Sukumar Periwal, 182–207. Budapest: Central European University Press.

Keating, Michael. 1997. "The Invention of Regions: Political Restructuring and Territorial Government in Western Europe." *Environment and Planning C: Government and Policy* 15:383–98.

Keck, Margaret E., and Kathryn Sikkink. 1998. *Activists beyond Borders: Advocacy Networks in International Politics*. Ithaca, NY: Cornell University Press.

Kellner, Douglas. 2002. "Theorizing Globalization." *Sociological Theory* 20 (3): 285–305.

Kepel, Gilles. 1987. *Les Banlieues de l'Islam*. Paris: Ed. du Seuil.

Kiefer, Markus. 1992. *Auf der Suche nach nationaler Identität und Wegen zur deutschen Einheit: Die deutsche Frage in der Überregionalen Tages–und Wochenpresse der Bundesrepublik, 1949–1955*. Frankfurt am Main: Peter Lang.

Kiez International. 1991. *Anlaufstellen und Projekte für Immigrantinnen, Immigranten und Flüchtlinge im Bezirk Kreuzberg*. Berlin: Senatsverwaltung für Soziales-Ausländerbeauftragte.

Kitschelt, Herbert. 1995. *The Radical Right in Western Europe*. Ann Arbor: University of Michigan Press.

Koenigsberger, Helmut Georg. 1986. "*Dominium Regale* or *Dominium Politicum et Regale*." In *Politicians and Virtuosi: Essays in Early Modern History*, by Helmut Georg Koenigsberger, 1–25. London: Hambleton.

Kohli, Martin. 2000. "The Battle-Grounds of European Identity." *European Societies* 2 (2): 113–37.

Koopmans, Ruud, and Paul Statham. 1999. "Challenging the Liberal Nation-State? Postnationalism, Multiculturalism, and the Collective Claims Making of Migrants and Ethnic Minorities in Britain and Germany." *American Journal of Sociology* 105 (3): 652–96.

Koselleck, Reinhard. 1988. *Critique and Crisis*. Cambridge: MIT Press.

Krasner, Stephen. 1999. *Sovereignty*. Princeton, NJ: Princeton University Press.

Kratochwil, Friedrich. 1986–87. "Of Systems, Boundaries, and Territoriality: An Inquiry in the Formation of the State System." *World Politics* 39 (1): 27–52.

Kraus, Peter. 2000. "Political Unity and Linguistic Diversity in Europe." *European Journal of Sociology* 41 (1): 138–63.

Kuzmics, Helmut, and Roland Axtmann. 2000. *Autorität, Staat und Nationalcharakter: Der Zivilisationsprozess in Österreich und England, 1700–1900*. Opladen, Germany: Leske und Budrich.

Kymlicka, William. 1995. *Multicultural Citizenship: A Liberal Theory of Minority Rights*. Oxford: Clarendon Press.

Laguerre, S. Michel. 1998. *Diasporic Citizenship: Haitian Americans in Transnational America*. New York: St. Martin's.

Laitin, David D. 1997. "The Cultural Identities of a European State." *Politics and Society* 25 (3): 277–302.

———. 1998. *Identity in Formation*. Ithaca, NY: Cornell University Press.

———. 2000. "Culture and National Identity: 'The East' and European Integration." EUI Working Paper, RSC no. 2000/3, European University Institute, Florence, Italy.

Lamont, Michele. 2000. *The Dignity of Working Men: Morality and the Boundaries of Race, Class, and Immigration*. Cambridge: Harvard University Press; New York: Russell Sage Foundation.

Lamont, Michele, and Laurent Thévénot, eds. 2000. *Rethinking Comparative Cultural Sociology: Repertoires of Evaluation in France and the United States*. New York: Cambridge University Press.

"La Padania ha bisogna di martiri." 1998. *Corriere della Sera* (Milan), 29 June, 5.

Lasch, Christopher. 1995. *The Revolt of the Elites and the Betrayal of Democracy*. New York: Norton.

Laurent, Yann. 2003. "Existe-t-il une 'mémoire commune européenne'?" *Le Monde*, March 5. http://lemonde.fr.

Leca, Jean. 1986. "Individualisme et citoyenneté." In *Sur l'individualisme*, ed. Pierre Birnbaum and Jean Leca, 159–213. Paris: Presses de la FNSP.

———. 1992. "Après Maastricht, sur la prétendu résurgence du nationalisme." *Témoins* 1 (1): 29–39.

Lefebvre, Henri. 1978. *De l'état: Les contradictions de l'état moderne*. Vol. 4. Paris: Union Générale d'Editions.

Lefèvre, Christian. 1998. "Metropolitan Government and Governance in Western Countries: A Critical Overview." *International Journal of Urban and Regional Research* 22 (1): 9–25.

Leitner, Helga, and Eric Sheppard. 1998. "Economic Uncertainty, Inter-Urban Competition, and the Efficacy of Entrepreneurialism." In Hall and Hubbard 1998, 285–308.

Lenoble, Jacques, and Nicole Dewandre, eds. 1992. *L'Europe au soir du siècle: Identité et démocratie*. Paris: Editions Espirit.

Leveau, Rémy. 1994. "Eléments de réflexion sur l'Islam en Europe." *Revue européenne des migrations internationales* 10 (1): 157–69.

Levitt, Peggy. 1998. "Local-Level Global Religion: The Case of the U.S.-Dominican Migration." *Journal for the Scientific Study of Religion* 37 (1): 74–89.

Lewis, Philip. 1994. *Islamic Britain: Religion, Politics, and Identity among British Muslims*. London: I. B. Tauris.

Lieven, Dominic. 2001. *Empire: The Russian Empire and Its Rivals*. New Haven, CT: Yale University Press.

Lipietz, Alain. 1994. "The National and the Regional: Their Autonomy vis-à-vis the Capitalist World Crisis." In *Transcending the State-Global Divide,* ed. R. Palan and B. Gills, 23–44. Boulder, CO: Lynne Rienner.

Lithman, Yngve Georg. 1997. "Spatial Concentration and Mobility." Paper presented to the Second International Metropolis Conference, Copenhagen, 25–28 September.

Livi-Bacci, Massimo. 2000. *The Population of Europe: A History*. Malden, MA: Blackwell.

Logan, John, and Harvey Molotch. 1987. *Urban Fortunes: The Political Economy of Place*. Berkeley: University of California Press.

Luke, Timothy W. 1989. *Screens of Power: Ideology, Domination, and Resistance in Informational Society*. Urbana: University of Illinois Press.

Lukes, Steven. 1973. *Individualism*. New York: Harper and Row.

Lustick, Ian S. 1993. *Unsettled States/Disputed Lands: Britain and Ireland, France and Algeria, Israel and the West Bank–Gaza*. Ithaca, NY: Cornell University Press.

———. 2000. "Agent-Based Modelling of Collective Identity: Testing Constructivist Theory." *Journal of Artificial Societies and Social Simulation* 3 (January) (1). jasss.soc.surrey.ac.uk/3/1.

Lustick, Ian S., and Dan Miodownik. 2000. "Deliberative Democracy and Public Discourse: The Agent-Based Argument Repertoire Model." *Complexity* 5 (4): 13–30.

———. 2002. "The Institutionalization of Identity: Micro Adaptation, Macro Effects, and Collective Consequences." *Studies in Comparative International Development* 37 (2) (summer): 24–53.

MacLeod, Gordon. 2000. "The Learning Region in an Age of Austerity: Capitalizing on Knowledge, Entrepreneurialism, and Reflexive Capitalism." *Geoforum* 31:219–36.

MacLeod, Gordon, and Mark Goodwin. 1999. "Space, Scale, and State Strategy: Rethinking Urban and Regional Governance." *Progress in Human Geography* 23 (4): 503–27.

Maier, Charles S. 2000. "Consigning the Twentieth Century to History: Alternative Narratives for the Modern Era." *American Historical Review* 105 (3): 807–31.

Malkki, Liisa. 1992. "National Geographic: The Rooting of Peoples and the Territorialization of National Identity among Scholars and Refugees." *Cultural Anthropology* 7:24–44.

Mandel, Ruth. 1996. "A Place of Their Own: Contesting Spaces and Defining Places in Berlin's Migrant Community." In *Making Muslim Space in North America and Europe,* ed. Barbara Daly Metcalf, 147–66. Berkeley: University of California Press.

Mann, Michael. 1993. "Nation-States in Europe and Other Continents: Diversifying, Developing, Not Dying." *Daedalus* 122 (3): 115–40.

———. 1997. "Has Globalization Ended the Rise and Rise of the Nation-State?" *Review of International Political Economy* 4 (autumn): 472–96.

———. 1998. "Is There a Society Called Euro?" In *Globalization and Europe,* ed. R. Axtmann, 184–207. London: Pinter.

Manners, Ian. 2001. "The 'Difference Engine': Constructing and Representing the International Identity of the European Union." Working Paper 40 in Copenhagen Peace Research Institute, Copenhagen.

Manners, Ian, and Richard Whitman. 1998. "Towards Identifying the International Identity of the European Union: A Framework for Analysis of the EU's Network of Relations." *Journal of European Integration* 21 (2): 231–49.

Mansbridge, Jane. 1998. "What Is the Public?" in *Private Action and the Public Good*, ed. W. Powell and E. Clemens. New Haven, CT: Yale University Press.

Mansell, Robin, and Roger Silverstone, eds. 1997. *Communication by Design: The Politics of Information and Communications Technologies*. Oxford: Oxford University Press.

Marshall, T. H. 1964. "Citizenship and Social Class." In *Class, Citizenship, and Social Development: Essays by T. H. Marshall*, 71–134. Chicago: University of Chicago Press.

Martin, Ron, and Peter Sunley. 1997. "The Post-Keynesian State and the Space Economy." In *Geographies of Economies*, ed. R. Lee and J. Wills, 278–89. London: Arnold.

Massey, Douglas, and Nancy Denton. 1993. *American Apartheid*. Cambridge: Harvard University Press.

Mauro, Ezio. 2001. "Un atto di fede diventa realta." *La Repubblica* (Rome), 31 December, 1.

Mayer, Margit. 1992. "The Shifting Local Political System in European Cities." In *Cities and Regions in the New Europe*, ed. M. Dunford and G. Kafkalas, 255–76. New York: Belhaven Press.

———. 1994. "Post-Fordist City Politics." In *Post-Fordism: A Reader*, ed. A. Amin, 316–37. Cambridge, MA: Blackwell.

McGuigan, Jim. 1996. *Culture and the Public Sphere*. London: Routledge.

McQuail, Dennis. 2001. "Media in Europe." In *Governing European Diversity*, ed. M. Buibernau. London: Sage.

Mehta, Uday Singh. 1999. *Liberalism and Empire*. Chicago: University of Chicago Press.

Merton, Robert K. 1957. *Social Theory and Social Structure*. New York: Free Press.

Miglio, Gianfranco, and Augusto Barbera. 1997. *Federalismo e secessione: Un dialogo*. Milan: Mondadori.

Miller, David. 1995. *On Nationality*. Oxford: Clarendon Press.

Milward, Alan S. 2000. *The European Rescue of the Nation-State*. 2d ed. London: Routledge.

Modood, Tariq. 1997. "Introduction: The Politics of Multiculturalism in the New Europe." In Modood and Werbner 1997, 1–25.

Modood, Tariq, and Pnina Werbner, eds. 1997. *The Politics of Multiculturalism in the New Europe: Racism, Identity, and Community*. London: Zed Books.

Mommsen, Wolfgang J., ed. 1994. *The Long Way to Europe: Historical Observations from a Contemporary View*. Chicago: Edition q.

Monahan, Anthony P. 1994. *From Personal Duties towards Personal Rights: Late Medieval and Early Modern Political Thought, 1300–1600*. Montreal: McGill-Queen's University Press.

Money, Jeannette. 1999. *Fences and Neighbors: The Political Geography of Immigration Control*. Ithaca, NY: Cornell University Press.

Moore, Sally F. 1989. "The Production of Cultural Pluralism as a Process." *Public Culture* 1:26–48.

Moravcsik, Andrew. 1998. *The Choice for Europe: Social Purpose and State Power from Messina to Maastricht.* Ithaca, NY: Cornell University Press.

Morris, Lydia. 1997. "Globalization, Migration, and the Nation-State: The Path to a Post-National Europe?" *British Journal of Sociology* 48 (2): 192–209.

Moscovici, Serge. 1984. "The Phenomenon of Social Representations." In *Social Representations,* ed. Robert M. Farr and Serge Moscovici, 3–71. New York: Cambridge University Press.

Mosse, Georg. 1975. *The Nationalization of the Masses.* Ithaca, NY: Cornell University Press.

Mouffe, Chantal. 1992. "Preface: Democratic Politics Today." In *Dimensions of Radical Democracy: Pluralism, Citizenship, Community,* ed. Chantal Mouffe, 1–14. London: Verso.

Mouritzen, Poul Erik. 1992. *Managing Cities in Austerity.* London: Sage.

Müftüler-Bac, Meltem. 1997. *Turkey's Relations with a Changing Europe.* Manchester: Manchester University Press.

Mukerji, Chandra. 1997. *Territorial Ambitions and the Gardens of Versailles.* Cambridge: Cambridge University Press.

Muldoon, James. 1999. *Empire and Order.* Houndmills, Basingstoke: Macmillan.

Nagel, Thomas. 1998. "Concealment and Exposure." *Philosophy and Public Affairs* 27:3–30.

Nairn, Tom. 2000. *After Britain.* London: Granta Books.

Nederman, Cary J., and Katie L. Forhan, eds. 1993. *Medieval Political Theory—A Reader: The Quest for the Body Politic, 1100–1400.* London: Routledge.

Nederveen Pieterse, Jan. 1991. "Fictions of Europe." *Race and Class* 32 (3): 3–10.

———. 1994. "Globalisation as Hybridisation." *International Sociology* 9 (2): 161–84.

Neumann, I. B., and J. Welsh. 1991. "The Other in European Self-Definition: An Addendum to the Literature on International Society." *Review of International Studies* 17:327–48.

Neunreither, Karlheinz, and Antje Wiener, eds. 2000. *European Integration after Amsterdam.* Oxford: Oxford University Press.

Neveu, Catherine. 1994. "Citoyenneté ou racisme en Europe: Exception et complémentarité britanniques." *Revue européenne des migrations internationales* 10 (1): 95–109.

Nohl, Arnd-Michael. 1996. *Jugend in der Migration: Türkische Banden und Cliquen in emprisicher Analyse.* Baltmannsweiler, Germany: Schneider Verlag Hohengehren.

Noiriel, Gérard. 1996. *The French Melting Pot: Immigration, Citizenship, and National Identity.* Trans. Geoffrey de Laforcade. Minneapolis: University of Minnesota Press.

Nozick, Robert. 1974. *Anarchy, State, and Utopia.* New York: Basic Books.

Nussbaum, Martha C., ed. 1996. *For Love of Country: Debating the Limits of Patriotism.* Boston: Beacon Press.

OECD (Organization for Economic Co-operation and Development). 1976. *Regional Problems and Policies in OECD Countries.* Paris: OECD.

———. 1992. *SOPEMI: Trends in International Migration: Continuous Reporting System on Migration.* Paris: OECD.

Offe, Claus. 1984. "'Crisis of Crisis Management': Elements of a Political Crisis Theory." In *Contradictions of the Welfare State,* 35–64. Cambridge: MIT Press.

———. 1996. "Modern 'Barbarity': A Micro-State of Nature?" *Constellations* 2 (3): 355–77.

———. 1998. "'Homogeneity' and Constitutional Democracy: Coping with Identity Conflicts through Group Rights." *Journal of Political Philosophy* 6:113–41.

O'Neill, Onora. 1996. *Towards Justice and Virtue: A Constructive Account of Practical Reason.* Cambridge: Cambridge University Press.

Orloff, Ann S. 1993. "Gender and the Social Rights of Citizenship." *American Sociological Review* 58 (3): 303–28.

Ossorio-Capella, Carles. 1972. *Der Zeitungsmarkt in der Bundesrepublik Deutschland.* Frankfurt am Main: Athenäum Verlag.

Owen, H. P. 1974. "Perfection." In *The Encyclopedia of Philosophy,* 5–6:87–88. New York: Macmillan.

Pajetta, Giovanna. 1994. *Il grande camaleonte: Episodi, passioni, avventure del leghismo.* Milan: Feltrinelli.

Parkinson, Michael. 1991. "The Rise of the Entrepreneurial European City: Strategic Responses to Economic Changes in the 1980s." *Ekistics* 350:299–307.

Peck, Jamie, and Adam Tickell. 1994. "Searching for a New Institutional Fix." In *Post-Fordism: A Reader,* ed. A. Amin, 280–315. Cambridge, MA: Blackwell.

Peterson, S. 1972. "Events, Mass Opinion, and Elite Attitudes." In *Communication in International Politics,* ed. R. L. Merritt. Urbana: University of Illinois Press.

Pickus, M. J. Noah. 1998. *Immigration and Citizenship in the 21st Century.* Rowmar, MD: Littlefield.

Pilkington, Hilary. 1994. *Russia's Youth and Its Culture: A Nation's Constructors and Constructed.* London: Routledge.

Pina-Cabral, João de, and John Campbell, eds. 1992. *Europe Observed.* London: Macmillan.

Pocock, J. G. A. 1997. "What Do We Mean by Europe?" *Wilson Quarterly* 31:12–29.

Pogge, Thomas. 1997. "Creating Supra-National Institutions Democratically: Reflections on the European Union's 'Democratic Deficit,'" *Journal of Political Philosophy* 5:163–82.

Poggi, Gianfranco. 1978. *The Development of the Modern State: A Sociological Introduction.* Stanford, CA: Stanford University Press.

Portes, Alexandro. 1996. "Transnational Communities: Their Emergence and Significance in the Contemporary World-System." In *Latin-America in The World-Economy,* ed. Roberto Patricio Korzeniewicz and William C. Smith, 151–66. Westport, CT: Greenwood Press.

Pred, Allan. 2000. *Even in Sweden: Racisms, Racialized Spaces, and the Popular Geographical Imagination.* Berkeley: University of California Press.

Press, Volker. 1997. *Das Alte Reich.* Berlin: Duncker and Humblot.

———. 1998. *Adel im Alten Reich.* Tübingen: Bibliotheca Academica Verlag.

Pressemitteilung. 1997. *Berliner Jugendliche türkischer Herkunft.* Berlin: Die Ausländerbeauftragte des Senats von Berlin.

Preuss, Ulrich K. 1998. "Citizenship in the European Union." In *Re-Imagining the Political Community,* ed. D. Archibugi, D. Held, and M. Köhler, 138–51. Stanford, CA: Stanford University Press.

Putnam, Robert D. 1993. *Making Democracy Work: Civic Traditions in Modern Italy*. Princeton, NJ: Princeton University Press.

Ram, Uri. 1998. "Postnational Pasts." *Social Science History* 22 (4): 513–45.

Rawls, John. 1971. *A Theory of Justice*. Cambridge: Harvard University Press.

Rex, John. 1994. "The Second Project of Ethnicity: Transnational Migrant Communities and Ethnic Minorities in Modern Multicultural Societies." *Innovation* 7 (3): 207–17.

———. 1995. "Ethnic Identity and the Nation State: The Political Sociology of Multi-Cultural Societies." *Social Identities* 1 (1): 21–34.

Reynolds, Susan. 1984. *Kingdoms and Communities in Western Europe, 900–1300*. New York: Oxford University Press.

Rhodes, Martin. 1995. "'Subversive Liberalism': Market Integration, Globalization, and the European Welfare State." *Journal of European Public Policy* 2 (3): 384–406.

Risse, Thomas. 2001. "A European Identity? Europeanization and the Evolution of Nation-State Identities." In *Transforming Europe*, ed. Maria Green Cowles, James Caporaso, and Thomas Risse, 198–216. Ithaca, NY: Cornell University Press.

Risse, Thomas, Daniela Engelmann-Martin, Hans-Joachim Knopf, and Klaus Roscher. 1999. "To Euro or Not to Euro? The EMU and Identity Politics in the European Union." *European Journal of International Relations* 5 (2): 147–87.

Rivera-Salgado, G. 1999. "Mixtec Activism in Oaxacalifornia: Transborder Grassroots Political Strategies." *American Behavioral Scientist* 42 (9): 1439–58.

Robbins, Keith. 1990. "National Identity and History: Past, Present, and Future." *History* 75:369–87.

Robins, Kevin, and Asu Aksoy. 1995. "Culture and Marginality in the New Europe." In *Europe at the Margins*, ed. Costis Hadjimichalis and David Sadler, 47–72. Chichester: Wiley.

Rodríguez-Pose, Andrés. 1998. *The Dynamics of Regional Growth in Europe*. Oxford: Clarendon Press.

Rokkan, Stein. 1975. "Dimensions of State Formation and Nation-Building: A Possible Paradigm for Research on Variations within Europe." In *The Formation of National States in Western Europe*, ed. Charles Tilly, 562–600. Princeton, NJ: Princeton University Press.

Romero, Federico. 1990. "Cross-Border Population Movements." In *The Dynamics of European Integration*, ed. William Wallace, 171–91. London: Pinter.

Ron, James. 2000. "Boundaries and Violence: Repertoires of State Action along the Bosnia/Yugoslavia Divide." *Theory and Society* 29 (5): 609–49.

Rosanvallon, Pierre. 2000. *The New Social Question: Rethinking the Welfare State*. Trans. Barbara Harshaw. Princeton, NJ: Princeton University Press.

Rosenau, James N. 1961. *Public Opinion and Foreign Policy: An Operational Formulation*. New York: Random House.

Ross, George. 1995. *Jacques Delors and European Integration*. Cambridge, Eng.: Polity Press.

Ruggie, John Gerard. 1993. "Territoriality and Beyond: Problematizing Modernity in International Relations." *International Organization* 47 (1): 139–74.

Rushdie, Salman. 1992. "In Good Faith." In *Imaginary Homelands*, by Salman Rushdie, 393–414. London: Granta Books.

Rusinow, Dennison. 1992. "Ethnic Politics in the Habsburg Monarchy and Successor States." In *Nationalism and Empire: The Habsburg Monarchy and the Soviet Union,* ed. Richard L. Rudolph and David F. Good, 243–67. New York: St. Martin's Press.

Sack, Robert David. 1986. *Human Territoriality: Its Theory and History*. Cambridge: Cambridge University Press.

———. 1997. *Homo Geographicus*. Baltimore: Johns Hopkins University Press.

———. 1999. "A Sketch of a Geographical Theory of Morality." *Annals of the Association of American Geographers* 89:26–44.

Sahlins, Peter. 1989. *Boundaries: The Making of France and Spain in the Pyrenees*. Berkeley: University of California Press.

Sandel, Michael. 1996. *Democracy's Discontent: America in Search of a Public Philosophy*. Cambridge: Harvard University Press.

Sassen, Saskia. 1996. *Losing Control? Sovereignty in an Age of Globalization*. New York: Columbia University Press.

———. 1998. *Globalization and Its Discontents*. New York: New Press.

———. 2000. "Territory and Territoriality in the Global Economy." *International Sociology* 15 (2): 372–93.

Saunders, Peter. 1979. *Urban Politics: A Sociological Interpretation*. London: Heinemann.

Scarry, Elaine. 1999. "The Difficulty of Imagining Other Persons." In *Human Rights in Political Transitions,* ed. Carla Hesse and Robert Post, 277–309. New York: Zone Books.

Schain, Martin A. 1987. "The National Front and the Construction of Political Legitimacy." *West European Politics* 10 (2): 229–52.

———. 1996. "The Immigration Debate and the National Front." In *Chirac's Challenge: Liberalization, Europeanization, and Malaise in France,* ed. Martin A. Schain and John T. S. Keeler, 169–97. New York: St. Martin's Press.

Scharpf, Fritz Wilhelm. 1999. *Governing in Europe: Effective and Democratic?* New York: Oxford University Press.

Schierup, Carl-Ulrik. 1995. "A European Dilemma: Myrdal, the American Creed, and EU Europe." *International Sociology* 10 (4): 347–67.

Schiffauer, Werner. 1991. *Die Migranten aus Subay: Türken in Deutschland, eine Ethnographie*. Stuttgart: Kleff-Cotta.

Schindling, Anton. 1991. *Die Anfänge des Immerwährenden Reichstags zu Regensburg*. Mainz: Verlag Philipp von Zabern.

Schlesinger, Philip. 1992a. "Europeanness: A New Cultural Battlefield?" *Innovation* 5 (1): 11–23.

———. 1992b. *Politics, Media, Identity*. London: Routledge.

———. 1997. "From Cultural Defense to Political Culture: Media, Politics, and Collective Identity in the European Union." *Media, Culture, and Society* 19 (3).

Schmidt, Georg. 1999. *Geschichte des Alten Reiches*. Munich: Beck.

Schmidt, Helmut. 2000. *Die Selbstbehauptung Europas: Perspektiven für das 21. Jahrhundert*. Stuttgart: Deutsche Verlags-Anstalt.

Schmitter, Philippe C. 2000. *How to Democratize the European Union . . . and Why Bother?* New York: Rowman and Littlefield.

———. 2001. "The Scope of Citizenship in a Democratized European Union: From Economic to Social and Cultural." In Eder and Giesen 2001, 86–131.

Schnapper, Dominique. 2002. "Citizenship and National Identity in Europe." *Nations and Nationalism* 8 (1): 1–14.

Schultz, E. 1984. "Unfinished Business: The German National Question and the Future of Europe." *International Affairs* 60 (3): 391–402.

Scott, Allen J. 1998. *Regions and the World Economy.* Oxford: Oxford University Press.

Sennett, Richard. 1977. *The Fall of Public Man.* New York: Norton.

Sewell, William H., Jr. 1991. "Introduction." *Social Science History* 16:3.

Seymour-Ure, Colin. 1968. *The Press, Politics, and the Public.* London: Methuen.

———. 1991. *The British Press and Broadcasting since 1945.* Oxford: Blackwell.

Sharpe, Laurence James. 1995. "The Future of Metropolitan Government." In *The Government of World Cities: The Future of the Metro Model,* ed. L. J. Sharpe, 11–31. New York: Wiley.

Shaw, David, Vincent Nadin, and Tim Westlake. 1996. "Towards a Supranational Spatial Development Perspective: Experience in Europe." *Journal of Planning Education and Research* 15:135–42.

Shin, Gi-Wook, James Freda, and Gihong Yi. 1999. "The Politics of Ethnic Nationalism in Divided Korea." *Nations and Nationalism* 5 (4): 465–84.

Shore, Cris. 2000. *Building Europe: The Cultural Politics of European Integration.* London: Routledge.

Shore, Cris, and Annabel Black. 1994. "Citizens' Europe and the Construction of European Identity." In Goddard, Llobera, and Shore 1994, 275–98.

Siedentop, Larry. 2000. *Democracy in Europe.* London: Penguin.

Silverman, Maxim. 1992. *Deconstructing the Nation: Immigration, Racism, and Citizenship in Modern France.* London: Routledge.

Skinner, Quintin. 1999. "Hobbes and the Purely Artificial Person of the State." *Journal of Political Philosophy* 7:1–29.

Smith, Anthony D. 1991. *National Identity.* London: Penguin.

———. 1992. "National Identity and the Idea of European Unity." *International Affairs* 68 (1): 55–76.

———. 1993. "A Europe of Nations—or the Nation of Europe." *Journal of Peace Research* 30 (2): 129–35.

———. 1995a. *Nations and Nationalism in a Global Era.* Cambridge, Eng.: Polity Press.

———. 1995b. "The Nations of Europe after the Cold War." In *Governing the New Europe,* ed. J. Hayward and E. C. Page, 44–66. Cambridge, Eng.: Polity Press.

———. 1996. "Culture, Community, and Territory: The Politics of Ethnicity and Nationalism." *International Affairs* 72 (3): 445–58.

———. 2000. *The Nation in History: Historiographical Debates about Ethnicity and Nationalism.* Hanover, NH: University Press of New England.

———. 2001. *Nationalism: Theory, Ideology, History.* Cambridge, Eng.: Polity Press.

Smith, Jackie, Charles Chatfield, and Ronald Pagnucco, eds. 1997. *Transnational Social Movements and Global Politics: Solidarity beyond the State.* Syracuse, NY: Syracuse University Press.

Smith, Karen. 2001. "The EU, Human Rights, and Relations with Third Countries: 'Foreign Policy' with an Ethical Dimension?" In *Ethics and Foreign Policy*, ed. K. Smith and M. Light. Cambridge: Cambridge University Press.

Snow, David A., and Robert D. Benford. 1992. "Master Frames and Cycles of Protest." In *Frontiers of Social Movement Theory*, ed. Aldon Morris and Carol McClurg Mueller, 133–56. New Haven, CT: Yale University Press.

Snow, David A., E. Burke Rochford Jr., Steven K. Worden, and Robert D. Benford. 1986. "Frame Alignment Processes, Micromobilization, and Movement Participation." *American Sociological Review* 51 (4): 464–82.

Soja, Edward W. 1989. *Postmodern Geographics*. London: Verso.

Somers, Margaret R. 1993. "Law, Community, and Political Culture in the Transition to Democracy." *American Sociological Review* 58 (5): 587–620.

———. 1994. "The Narrative Constitution of Identity: A Relational and Network Approach." *Theory and Society* 23:605–49.

Soysal, Levent. 1999. "Projects of Culture: An Ethnographic Episode in the Life of Migrant Youth in Berlin." Ph.D. diss., Harvard University.

———. 2001. "Diversity of Experience, Experience of Diversity: Turkish Migrant Youth Culture in Berlin." *Cultural Dynamics* 13 (1): 5–28.

Soysal, Yasemin Nuhoğlu. 1994. *Limits of Citizenship: Migrants and Postnational Membership in Europe*. Chicago: University of Chicago Press.

———. 1997. "Changing Parameters of Citizenship and Claims-Making: Organized Islam in European Public Spheres." *Theory and Society* 26:509–27.

———. 2002. "Locating Europe." *European Societies* 4 (3): 265–84.

Speir, Hans. 1941. "Magic Geography." *Social Research* 8:310–30.

Spillman, Lyn. 1997. *Nation and Commemoration*. Cambridge: Cambridge University Press.

Spruyt, Hendrik. 1994. *The Sovereign State and Its Competitors*. Princeton, NJ: Princeton University Press.

Statistical Office of the European Communities. 1992. *Labour Force Survey*. Luxembourg: Office for Official Publications of European Communities.

Stöhr, Walter, and D. R. Fraser Taylor. 1981. *Development from Above or Below? The Dialectics of Regional Planning in Developing Countries*. New York: Wiley.

Stolcke, Verena. 1995. "Talking Culture: New Boundaries, New Rhetorics of Exclusion in Europe." *Current Anthropology* 36 (1): 1–24.

Strath, Bo, ed. 2000. *Europe and the Other and Europe as the Other*. Brussels: Peter Lang.

Streeck, Wolfgang. 2001. "Citizenship under Regime Competition: The Case of European 'Works Councils.'" In Eder and Giesen 2001, 122–56.

Sudre, Frederic. 1995. *Droit international et européen des droits de l'homme*. Paris: Presse Universitaire Français.

Sunstein, Cass. 1993. *The Partial Constitution*. Cambridge: Harvard University Press.

Suny, Ronald Grigor. 2001. "Constructing Primordialism: Old Histories for New Nations." *Journal of Modern History* 73:862–96.

Sweet, Alec Stone, Neil Fligstein, and Wayne Sandholtz. 2001. "The Institutionalization of European Space." In *The Institutionalization of Europe,* ed. Alec Stone Sweet, Neil Fligstein, and Wayne Sandholtz, 1–28. New York: Oxford University Press.

Swidler, Ann. 1986. "Culture in Action: Symbols and Strategies." *American Sociological Review* 51:273–86.

Swyngedouw, Erik. 2000. "Authoritarian Governance, Power, and the Politics of Rescaling." *Environment and Planning D: Society and Space* 18:63–76.

Tambini, Damian. 2001. "Post-National Citizenship." *Ethnic and Racial Studies* 24 (2): 195–217.

Tarrow, Sidney. 2001. "La contestation transnationale." *Cultures et conflits* 38–39:187–225.

———. Forthcoming. "Center-Periphery Alignments and Political Contention in Late-Modern Europe." In *Restructuring Territoriality: Europe and the United States Compared,* ed. Christopher K. Ansell and Giuseppe DiPalma. Cambridge: Cambridge University Press.

Tassin, Etienne. 1992. "Europe: A Political Community." In *Dimensions of Radical Democracy: Pluralism, Citizenship, Community,* ed. C. Mouffe, 169–92. London: Verso.

Taylor, Charles. 1989. *Sources of the Self.* Cambridge: Harvard University Press.

———. 1992. "Les institutions dans la vie nationale." In *Rapprocher les solitudes: Ecrits sur le fédéralisme et le nationalisme au Canada,* ed. Guy Laforest, 135–53. Sainte-Foy, Quebec: Presses de l'Université de Laval.

———. 1995. *Philosophical Arguments.* Cambridge: Harvard University Press.

———. 2002. "Modern Social Imaginaries." In Gaonkar 2002.

Taylor, Peter J. 1994. "The State as Container: Territoriality in the Modern World-System." *Progress in Human Geography* 18 (2): 151–62.

Teitelbaum, Michael S., and Jay Winter. 1998. *A Question of Numbers: High Migration, Low Fertility, and the Politics of National Identity.* New York: Hill and Wang.

Tertilt, Hermann. 1996. *Turkish Boys: Ethnographie einer Jugendbande.* Frankfurt am Main: Suhrkamp Verlag.

Therborn, Goran. 2001. "European Modernity and European Normativity: The EU in History and in Social Space." In *Institutional Approaches to the European Union: Proceedings from Arena Workshop,* ed. Svein S. Andersen.

Thomas, Elaine. 1998. "The New Cultural Politics of Muslim Minority Integration in France and Britain: Beyond Left and Right?" Paper presented to the Eleventh International Conference of Europeanists, Baltimore, 26–28 February.

Tilly, Charles. 1974. *Nation-State Formation: Reflections on the History of European State Making.* Princeton, NJ: Princeton University Press.

———. 1990. *Coercion, Capital, and European States, AD 990–1990.* Oxford: Blackwells.

———. 1994. "Entanglements of European Cities and States." In *Cities and the Rise of States in Europe: AD 1000 to 1800,* ed. Charles Tilly and Wim Blockman, 1–27. Boulder, CO: Westview Press.

Tocqueville, Alexis de. [1835] 1990. *Democracy in America.* Ed. Daniel J. Boorstin. New York: Vintage Books.

Torpey, John C. 2000. *The Invention of the Passport.* Cambridge: Cambridge University Press.

Torregrosa, José Ramón. 1993. "In Search of Hispanic Cultural Identity." Paper presented at the World Values Study Conference, 7–15 September, El Paular, Spain.

Trubowitz, Peter. 1998. *Defining the National Interest: Conflict and Change in American Foreign Policy.* Chicago: University of Chicago Press.

Tsing, Anna. 2000. "Inside the Economy of Appearances." *Public Culture* 12 (1): 115–44.

Tuan, Yi-Fu. 1977. *Space and Place: The Perspective of Experience.* Minneapolis: University of Minnesota Press.

———. 1989. *Morality and Imagination: Paradoxes of Progress.* Madison: University of Wisconsin Press.

———. 1996. *Cosmos and Hearth: A Cosmopolite's Viewpoint.* Minneapolis: University of Minnesota Press.

Tunstall, Jeremy. 1996. *Newspaper Power: The New National Press in Britain.* Oxford: Clarendon Press.

Turner, Bryan S. 2001. "The Erosion of Citizenship." *British Journal of Sociology* 52 (2): 189–209.

Turner, Victor. 1974. *Dramas, Fields, and Metaphors: Symbolic Action in Human Society.* Ithaca, NY: Cornell University Press.

United Nations. 2000. *Replacement Migration: Is It a Solution to Declining and Aging Population?* Population Division, Department of Economic and Social Affairs. New York: United Nations Press.

Varenne, Hervé. 1993. "The Question of European Nationalism." In Wilson and Smith 1993, 223–40.

Venturelli, Shalini. 1998. *Liberalizing the European Media: Politics, Regulation, and the Public Sphere.* Oxford: Clarendon.

Verdun, Amy. 1999. "The Logic of Giving Up National Currencies: Lessons from Europe's Monetary Union." In *Nation-States and Money: The Past, Present, and Future of National Currencies,* ed. Emily Gilbert and Eric Helleiner, 199–214. London: Routledge.

Vernet, Daniel. 1992. "The Dilemma of French Foreign Policy." *International Affairs* 68 (4): 655–64.

Vertovec, Steven. 1996. "Berlin Multikulti: Germany, 'Foreigners,' and 'World-Openness.'" *New Community* 22 (3): 381–99.

Viroli, Maurizio. 1995. *For Love of Country: An Essay on Patriotism and Nationalism.* New York: Clarendon Press.

Voltmer, Katrin. 1998. *Medienqualität und Demokratie.* Baden-Baden: Nomos.

Wacquant, Loïc J. D. 1996. "The Rise of Advanced Marginality: A Note on Its Nature and Implications." *Acta Sociologica* 39 (20): 121–39.

Waever, Ole. 1997. "The Baltic Sea: A Region after Post-Modernity?" In *Neo-Nationalism or Regionality: The Restructuring of Political Space around the Baltic Rim,* ed. P. Joenniemi, 293–342. Copenhagen: NordREFO.

———. 2000. "The EU as a Security Actor: Reflections from a Pessimistic Constructivist on Post-Sovereign Security Orders." In *International Relations Theory and the Politics of European Integration,* ed. M. Kelstrup and M. Williams, 250–94. London: Routledge.

Wagner-Pacifici, Robin. 2001. "Prolegomena to a Paradigm: Narratives of Surrender." *Qualitative Sociology* 24 (2): 269–81.

Walker, Neil. 1998. "Sovereignty and Differentiated Integration in the European Union." *European Law Journal* 4:355–88.

Walker, R. B. J. 1993. *Inside/Outside: International Relations as Political Theory*. New York: Cambridge University Press.

Wallace, Martin. 1986. *Recent Theories of Narrative*. Ithaca, NY: Cornell University Press.

Wallace, William. 1986. "What Price Independence? Sovereignty and Interdependence in British Politics." *International Affairs* 62 (3): 367–89.

———. 1994. "Rescue or Retreat? The Nation State in Western Europe, 1945–93." *Political Studies* 42:52–76.

———. 1999. "The Sharing of Sovereignty: The European Paradox." *Political Studies* 57:503–21.

Wallman, Sandra. 1978. "The Boundaries of 'Race': Processes of Ethnicity in England." *Man* 13:200–217.

Walzer, Michael. 1994. *Thick and Thin: Moral Argument at Home and Abroad*. Notre Dame, IN: Notre Dame University Press.

Warner, Michael. 2002. *Publics and Counterpublics*. Cambridge, MA: Zone Books.

Watkins, Susan Cott. 1991. *From Provinces into Nations*. Princeton, NJ: Princeton University Press.

Weale, Albert. 1995. "From Little England to Democratic Europe?" *New Community* 21 (2): 215–25.

Weber, Eugen. 1976. *Peasants into Frenchmen*. Stanford, CA: Stanford University Press.

Weber, Max. [1922] 1978. *Economy and Society: An Outline of Interpretive Sociology*. Ed. Gunther Roth and Claus Wittich. Berkeley: University of California Press.

Weil, Patrick. 2001. "The History of French Nationality: A Lesson for Europe." In *Towards a European Nationality: Citizenship, Immigration, and Nationality Law in the European Union*, ed. Randal Hansen and Patrick Weil, 52–68. New York: St. Martin's.

Weiler, Joseph H. 1999a. *The Constitution of Europe: "Do the New Clothes Have an Emperor?" and Other Essays on European Integration*. Cambridge: Cambridge University Press.

———. 1999b. "To Be a European Citizen: Eros and Civilization." in Weiler 1999a, 324–57.

Werbner, Pnina. 1997. "Afterword: Writing Multiculturalism and Politics in the New Europe." In Modood and Werbner 1997, 261–67.

Wessels, Bernhard. 1995. "Support for Integration: Elite or Mass-Driven." In *Public Opinion and Internationalized Governance*, ed. Richard Sinnott and Oskar Niedermayer, 137–63. Oxford: Oxford University Press.

White, Harrison. 1992. *Identity and Control: A Structural Theory of Social Action*. Princeton, NJ: Princeton University Press.

White, Jenny B. 1997. "Turks in the New Germany." *American Anthropologist* 99 (4): 754–69.

Whitman, Richard. 1998. *From Civilian Power to Superpower? The International Identity of the European Union*. Basingstoke: Macmillan.

Wieviorka, Michel. 1998. "Is Multiculturalism the Solution?" *Ethnic and Racial Studies* 21 (5): 881–910.

———, ed. 1992. *La France Raciste*. Paris: Seuil.

Willis, Paul. 1977. *Learning to Labor.* New York: Columbia University Press.

———. 1990. *Common Culture: Symbolic Work at Play in Everyday Cultures of the Young.* Boulder, CO: Westview Press.

Wilson, Peter. 1999. *The Holy Roman Empire, 1495–1806.* Houndmills, Basingstoke: Macmillan.

Wilson, Thomas M., and Hastings Donnan, eds. 1998. *Border Identities: Nation and State at International Frontiers.* Cambridge: Cambridge University Press.

Wilson, Thomas M., and M. Estellie Smith, eds. 1993. *Cultural Change and the New Europe: Perspectives on the European Community.* Boulder, CO: Westview Press.

Wolbert, Barbara. 1995. *Der getötete Pass: Rückkehr in die Türkei.* Berlin: Akademia Verlag.

Wolton, Dominique. 1998. "La communication et l'Europe: Du multiculturalisme à la cohabitation culturelle. In Kastoryano 1998, 65–81.

Wouters, Jan. 2000. "National Constitutions and the European Union." *Legal Issues of Economic Integration* 27 (1): 25–92.

Yapp, M. E. 1992. "Europe in the Turkish Mirror." *Past and Present* 137:134–55.

Zerubavel, Eviatar. 1992. *Terra Incognita: The Mental Discovery of America.* New Brunswick, NJ: Rutgers University Press.

Zincone, Giovanna. 1997. "The Powerful Consequences of Being Too Weak: The Impact of Immigration on Democratic Regimes." *Archives européennes de sociologie* 38 (1): 104–38.

Contributors

JOHN AGNEW is professor of geography at University of California at Los Angeles, where he teaches political geography. He is the author of *Place and Politics in Modern Italy* and *Making Political Geography* and a coeditor of *American Space/American Place: Geographies of the Contemporary United States*.

ROLAND AXTMANN is reader in politics and international relations and associate director of the Centre for the Study of Globalization at the University of Aberdeen, Scotland. Recent publications include *Liberal Democracy into the 21st Century, Globalization and Europe, Autoritaet, Staat und Nationalcharakter: Der Zivilisationsprozess in Oesterreich und England, 1700–1900, Balancing Democracy,* and *Understanding Democratic Politics*.

MABEL BEREZIN is associate professor of sociology at Cornell University. She is the author of *Making the Fascist Self: The Political Culture of Inter-War Italy,* winner of the 1998 J. David Greenstone Prize of the American Political Science Association for the best book in politics and history. Recent publications include *Democratic Culture: Ethnos and Demos in Global Perspective,* edited with Jeffrey C. Alexander; "Secure States: Towards a Political Sociology of Emotion," in *Emotions and Sociology;* and "Western European Studies: Culture," in *International Encyclopaedia of the Social and Behavioral Sciences*.

NEIL BRENNER is assistant professor of sociology and metropolitan studies at New York University. He has edited *Spaces of Neoliberalism: Urban Restructuring in North America and Western Europe* (with N. Theodore) and *State/Space: A Reader* (with B. Jessop, M. Jones, and G. Macleod).

CRAIG CALHOUN is president of the Social Science Research Council in New York City and professor of sociology and history at New York University. His most recent books include *Nationalism, Neither Gods nor Emperors: Students and the Struggle for Democracy in China,* and *Critical Social Theory*. He is the editor-in-chief of the *Oxford Dictionary of the Social Sciences* and coeditor for international and area studies of the *International Encyclopedia of Social and Behavioral Sciences*.

JUAN DÍEZ-MEDRANO is professor of integrated social sciences at International University, Bremen. He is the author of *Divided Nations: Class Conflict, Politics, and Nationalism in the Basque Country and Catalonia;* "Nested Identities and European Identity in Spain," *Ethnic and Racial Studies* 24 (2001): 5; and "The European Union: Economic Giant, Political Dwarf?" in *International Order and the Future of World Politics,* edited by T. V. Paul and John A. Hall, among other publications.

ROY J. EIDELSON is executive director of the Solomon Asch Center for Study of Ethnopolitical Conflict at the University of Pennsylvania.

J. NICHOLAS ENTRIKIN is professor of geography at the University of California, Los Angeles. He is also a member of the Executive Committee of the UCLA Institute for the Environment. Recent publications include *The Betweenness of Place.*

RIVA KASTORYANO is a senior research fellow at the CNRS (National Center for Scientific Research) and teaches at the Institute for Political Studies in Paris. Her most recent books include *La France, l'Allemagne et leurs Immigrés: Négocier l'identité,* translated into English as *Negotiating Identities: States and Immigrants in France and Germany.* She is the editor of *Quelle identité pour l'Europe? Le multiculturalisme à l'épreuve.*

KRISHAN KUMAR is the William R. Kenan Professor of Sociology at the University of Virginia. Among his publications are *Prophecy and Progress, Utopia and Anti-Utopia in Modern Times, From Post-Industrial to Post-Modern Society, 1989: Revolutionary Ideas and Ideals,* and *The Making of English National Identity.*

IAN S. LUSTICK is professor of political science and holds the Merriam Term Chair in Political Science at the University of Pennsylvania. His current research focuses on Israeli-Palestinian relations and on the development of advanced agent-based computer simulation techniques for studying comparative politics, with special emphasis on the Middle East.

MARTIN SCHAIN is professor of politics and director of the Center for European Studies at New York University. He is the coeditor of seven volumes, including most recently *A Century of Organized Labor in France* (with Herrick Chapman and Mark Kesselman) and *Shadows over Europe: The Development and Impact of the Extreme Right in Europe* (with Aristede Zolberg), as well as the coauthor of a widely diffused text on France, *Politics in France.*

LEVENT SOYSAL is a postdoctoral research fellow at the Berlin Program for Advanced German and European Studies, Free University–Berlin. His dissertation, titled "Projects of Culture: An Ethnographic Episode in the Life of Migrant Youth in Berlin," explores the project(s) of youth culture undertaken and staged by (and for) Turkish migrant youths in the public spaces of Berlin.

Index

Page numbers in *italics* refer to illustrations.